Procurement Systems

Also available from E & FN Spon

Building International Construction Alliances
R. Pietroforte

Construction – Craft to Industry
G. Sebestyen

Construction Methods and Planning
J.R. Illingworth

Creating the Built Environment
L. Holes

Green Building Handbook
T. Woolley, S. Kimmins, R. Harrison and P. Harrison

Industrialized and Automatic Building Systems
A. Warszawski

Introduction to Eurocode 2
D. Beckett and A. Alexandrou

Open and Industrialised Building
A. Sarja

The Idea of Building
S. Groak

Journal
Building Research and Information
**The International Journal of Research, Development, Demonstration &
Innovation**

*To order or obtain further information on any of the above or receive a full
catalogue please contact*:
The Marketing Department, E & FN Spon, 11 New Fetter Lane, London
EC4P 4EE.
Tel: 0171 842 2400; Fax: 0171 842 2303

Procurement Systems

A guide to best practice in construction

Edited by
Steve Rowlinson and
Peter McDermott

International Council for
Building Research Studies
and Documentation

First published 1999
by E & FN Spon
11 New Fetter Lane, London EC4P 4EE

Simultaneously published in the USA and Canada
by Routledge
29 West 35th Street, New York, NY 10001

E & FN Spon is an imprint of the Taylor & Francis Group

© 1999 E & FN Spon

Typeset in Sabon by J&L Composition Ltd, Filey, North Yorkshire
Printed and bound in Great Britain by St. Edmundsbury Press, Bury
St. Edmunds, Suffolk

British Library Cataloguing in Publication Data
A catalogue record for this book is available from the British
Library

Library of Congress Cataloging in Publication Data

ISBN 0-419-24100-0

Contents

List of contributors

Richard Fellows works as an independent construction management consultant and researcher in Hong Kong. Previously he was reader in Construction Management at the University of Bath, UK. He is joint coordinator of the CIB Task Group (TG-23)on Culture in Construction. Richard has published a wide variety of material particularly in the areas of construction economics and finance, contracts and law in books, refereed journals and conferences. He has worked as construction management coordinator for the UK's Engineering and Physical Sciences Research Council. Richard's current research involves various facets of culture – its impact on quality, on project manager's activities and on goals for construction projects.

Dr Stuart Green is a senior lecturer within the Department of Construction Management and Engineering at the University of Reading, UK. He is course director for the M.Sc. course in project management and also leads the Value Management Research Group. Stuart has a B.Sc. in civil engineering from the University of Birmingham, UK, a M.Sc. in Construction Management from Heriot-Watt University, UK and a Ph.D. from the University of Reading, UK. He has published extensively on the topics of value management, strategic briefing and participative process modelling. In addition to his research activities, Stuart also regularly acts as a consultant to industry. Experience includes value management, partnering and participative strategic planning.

Cliff Hardcastle holds three degrees including an M.Sc. by Research from the University of Strathclyde and a Ph.D. from Heriot Watt University, Edinburgh. Currently Head of Department of Building and Surveying at Glasgow Caledonian University, he was previously Head of Department at the University of Westminster.

Cliff's previous research work in construction procurement includes the potential of integrated databases for information exchange in the construction industry. During the late 1980's he was the main grant holder for a number of research council funded projects, all of which involved close collaboration with industry. These included the analysis of approaches to procurement and cost control of petrochemical works. He is currently co-investigator with Professor Langford researching the re-engineering of the building procurement decision making process. He also holds a major grant together with Strathclyde University for the establishment of a Centre of Advanced Built Environment Research to link researchers at the Glasgow Caledonian University, Strathclyde University and the Building Research Establishment (Scotland). He is an active member of the RICS Research Committee and of CIB W55 and has been on a number of conferences' scientific committees.

Bob Hindle is a senior lecturer and director of the BSc. Construction Management programme in the Department of Construction Economics and Management at the University of Cape Town, South Africa. He has been a senior lecturer for the past ten years prior to which he had a successful career in the construction industry where he held several senior positions, including that of CEO of a large construction contracting firm. He is a co-founder of the African Centre for Strategic Studies in Construction which is active in research concerned with construction industry development issues, relevant to the policy development process that is unfolding in South Africa at present. Much of his work has been concerned with procurement issues and construction industry development.

Mohan Kumaraswamy is an associate professor at the Department of Civil Engineering of the University of Hong Kong. Having earned a degree in Civil Engineering from the University of Peradeniya, Sri Lanka, in 1976, and worked in construction organisations in Sri Lanka and Nigeria, he later obtained an M.Sc. in construction management and then a Ph.D. from Loughborough University, UK, in 1991. The latter focused on evaluation systems for the management of construction projects.

His present research interests centre on construction procurement systems (including decision support for their selection and procurement sub-systems such as in contractor selection); technology exchange or transfers; construction claims, dispute resolution and avoidance; and construction planning and productivity enhancement issues. He has directed the development of construction management training packages for industry in Sri Lanka and is presently contributing to the development of a computer-aided teaching/learning package for civil engineering undergraduates.

Apart from his extensive industry experience, for example as a director of the first construction project management company in Sri Lanka in the 1980s, he has also been a senior consultant and team leader on many internationally funded projects (of the World Bank, United Nations Development Programme, International Labor Office and United Nations). He is active on councils and committees of many professional bodies, such as the Hong Kong Institution of Engineers, the Chartered Institute of Building and the Asian Construction Management Association and (previously) the Institution of Engineers Sri Lanka. He has been the vice chairman of the editorial board of the *Asia Pacific Building* and *Construction Management Journal* since its launch in 1995.

David Langford holds the Barr Chair of Construction at Strathclyde University in Glasgow, UK. He has worked in the Department of Building Technology (Brunel University 1976–87), Architecture and Building Engineering (Bath University 1987–91), and Civil Engineering (Strathclyde University 1991–present). His primary academic interests are in construction management and, in particular, strategic management, human resource management and project management applied to the construction industry.

He has published widely and books to which he has contributed include: *Construction Management in Practice*, *Direct Labour Organisations in Construction*, *Construction Management Vol. I* and *Vol. II*, *Strategic Management in Construction*, *Human Resource Management in Construction* and *Managing Overseas Construction*. He has contributed to seminars on the field of construction management in all five continents.

Professor Dennis Lenard is a Professor at the University of Technology, Sydney, specialising in Construction and Property. He has the following academic qualifications: Bachelor of Applied Science (BAppSc), with First Class Honours, Faculty of Architecture and Building, NSWIT currently University of Technology, Sydney; Master of Applied Science (MAppSc), Faculty of Architecture and Building, NSWIT currently University of Technology, Sydney, Thesis entitled "*Competitive Tendering Strategy for the Risk-Averse Contractor*"; Doctor of Philosophy (Ph.D.), Faculty of Architecture, University of Newcastle, NSW Australia. His thesis is entitled *Innovation and Industrial Culture in the Australian Construction Industry: A Comparative Benchmarking Analysis of the Critical Cultural Indices Underpinning Innovation*.

Professor Lenard is currently Director of the Construction Industry Institute, Australia which has been in operation since 1992. He is also a director of the newly formed Centre for Built Infrastructure Research a collaborative effort between three Faculties (Engineering, Science and Design, Architecture and Building) at the University of Technology, Sydney. He has obtained research support of more than A$3.5 m from a combination of industry and government grants. His current research is in Information Technology and in developing an intranet strategy for groups of industry participants.

Professor Lenard has held many senior positions in the Profession and Industry and is currently Senior Vice Chairman of the International Cost Engineering Council. He is also Director of Research of the Construction Industry Institute, Australia.

Professor Lenard is author and co-author of over 100 scholarly publications in Construction and Property.

Anita Liu has 12 years of teaching experience and is currently teaching in the Department of Real Estate and Construction, the University of Hong Kong. She has previous working experience in the construction industry with the quantity surveying consultant, the contractor and the government. Anita was the past chairperson of the Quantity Surveying Division, Royal Institute of Chartered Surveyors RICS (Hong Kong Branch) and the Hong Kong Institute of Surveyors. She is a member of the Surveyors Registration Board in Hong Kong. Her Ph.D. is in project outcome evaluation and her current research interests include project leadership, value management and conflict management.

Beverley Lloyd-Walker holds qualifications in business, education and information management. A lecturer in the Department of Management, Faculty of Business, Victoria University of Technology for eight years, Beverley has lectured in human resource management, training and change management and organisational change and development at undergraduate and postgraduate levels.

Current research areas include facilitation of major change programmes, impact of change on remaining staff, technology-supported change, information technology (IT) to support improved service quality, use of organisational knowledge for competitive advantage, and the use of human resource (HR) information systems to support devolution of HR activities. For her Ph.D., Beverley researched the contribution of IT to organisational change and performance in the Australian banking industry.

Jason Matthews is currently employed in the Department of Real Estate and Construction of the University of Hong Kong as a post-doctoral research fellow. At present he is developing a number of research initiatives with international contractors into partnering and strategic planning within Hong Kong and China.

Previous to this he worked as a management consultant for a leading Asia-Pacific construction management consultancy. He worked on a number of assignments including: preparing tender documents for Chep Lap Kok; coordinating planning proposals for the redevelopment of Kai Tak airport; and undertaking a productivity review of technical officers within the Architectural Services Department of the Hong Kong SAR Government.

Jason also has experience working for regional and international contractors in the United Kingdom. It was whilst employed at HBG Kyle Stewart that Jason commenced his doctoral research at Loughborough University into how partnering relationships can improve the main contractor–subcontractor relationship. His research was undertaken simultaneously with HBG Kyle Stewart implementing partnering approaches with their clients and subcontractors and suppliers.

His current research interests include procurement, partnering and collaboration, and strategic management, with particular emphasis on strategic planning.

Dr Peter McDermott holds an honours degree in quantity surveying and a doctorate in construction management. He is a principal researcher in the Research Centre for the Built and Human Environment (BUHU) at the University of Salford. BUHU achieved the highest possible research rating in the most recent UK government research assessment exercise. Over the past 12 years Peter has published extensively in the field of construction procurement and construction industry development. Currently he is involved in research projects on partnering, supply-chain management, lean construction, process management and the strategic management of contracting and professional firms.

Within the Department of Surveying at Salford University Peter applies this research to the teaching of management and procurement on undergraduate and postgraduate programmes in surveying and construction management. Peter is a founder member of the International Building Research Council, Working Group W92 – Procurement systems.

Dr John B. Miller is Associate Professor of Civil and Environmental Engineering at Massachusetts Institute of Technology (M.I.T.). His teaching and research activities are focused on improving the methods by which governments acquire infra-structure facilities and services, including public–private programme planning and procurement strategies. He is the principal author of numerous case studies related to public–private project delivery and finance of infrastructure. For 15 years prior to returning to M.I.T., he practised construction and government contracts law with Gadsby & Hannah in Boston and Washington, as an associate, general partner, and a managing partner. From 1994 to 1995, he served as Chair of the American Bar Association's Section of Public Contract Law. John holds a B.Sc. in Civil Engineering and an M.Sc. in soil mechanics from M.I.T., both in 1974. He also holds Juris Doctor (1977) and Master of Law (Taxation) (1982) degrees from Boston University School of Law. He received his Ph.D. in infrastructure development systems from M.I.T. in 1995.

Mike Murray is a Research Assistant in the Department of Civil Engineering in Strathclyde University, Glasgow, UK. He graduated from Napier University (in Edinburgh), with a First Class Honours Degree in Building Engineering & Management. He then continued his academic studies with an M.Sc. in Construction Management. He has lectured in Construction Management at the Robert Gordon University, Aberdeen and Heriot–Watt University, Edinburgh.

His current research interests involve the examination of organisational structures within building projects. Previous research includes investigation of trade skill shortages, within the UK Construction Industry. He has also conducted research into the opportunities available overseas for UK contractors and consultants.

George Norval graduated with a Bachelor of Science in Quantity Surveying degree in 1965, and a Bachelor of Commerce degree in 1969, both from the University of Natal, an institution at which he has lectured for the past twelve years after serving as managing director of a major Durban based construction company. He is a senior lecturer and is the Programme Director of Quantity Surveying. He is a Fellow of the Royal Institution of Chartered Surveyors and an alternate member of the Council of South African Quantity Surveyors.

He has served as external examiner for courses conducted at the Natal Technikon and the University of Cape Town, and is currently a Royal Institution of Chartered Surveyors examiner of the Quantity Surveying programme at the University of Port Elizabeth. He serves on the Boards of two University of Natal NGO's; 'The Built Environment Support Group', and the 'Peace Foundation', achieving considerable success nationally in the provision of school and civic buildings in depressed rural areas. He has had extensive experience dating back to 1987 working on community based projects and was instrumental in the formation of the 'Maputaland Builders Association' in 1993. His efforts in this remote segment of KwaZulu Natal have resulted in several major contracts being awarded to and completed by joint venture teams of emergent local builders, with components produced by local community suppliers.

Dr Steve Rowlinson is senior lecturer in the Department of Real Estate and Construction at the University of Hong Kong. Steve has worked in Hong Kong for 12 years and his research interests include procurement systems, visualisation and construction site safety. He has been co-coordinator of the W92 Procurement Systems Working Commission for the past six years and has written and published extensively in the area for procurement systems.

He obtained his first degree in civil engineering at the University of Nottingham, UK and achieved his Ph.D. whilst studying at Brunel University. He has been highly active in the W92 Working Commission and has been instrumental in organising many of the symposia and conferences which the Commission has run. During his time as co-coordinator for the Commission it has met at many locations around the world and has encompassed themes that reflect the growing diversity of the Commission's members interests and backgrounds.

Dr Pantaleo Rwelamila is an associate professor in the Department of Construction Economics and Management at the University of Cape Town, South Africa. He was formerly the director of engineering programmes and director of research in the Department of Civil Engineering, Faculty of Engineering and Technology, University of Botswana; Lecturer and Researcher at the School of Environmental Studies, University of Zambia at Ndola, Kitwe in Zambia and later Copper Belt University and consultant to SIDA (Sweden). Prior to his joining the University of Zambia, he was a principal quantity surveyor/cost engineer with the Department of Civil Engineering, Tanzania Railways Corporation.

Rob Taylor completed a B.Sc.(QS) degree (with distinction) at the University of Natal, Durban in 1976, thereafter registering as a quantity surveyor. He was elected a Professional Associate of the RICS in 1978. He proceeded to complete an M.Sc (Construction Management) degree at Heriot-Watt University in 1980. He lectured at the University of The Witwatersrand, Johannesburg, later moving to the University of Natal, Durban until 1987.

He then held the post of project manager in the Informal Settlements Division of the Urban Foundation, Natal regional office. He was involved in the upgrading of informal settlements around Durban, becoming manager of the unit in 1989.

In 1994 he became Professor of Construction Management at the University of Natal and was head of department until 1996.

A parallel interest in systems thinking, organisational learning and leadership ultimately led to the transfer in 1999, to the Faculty of Management Studies and the post of Acting Director of the new University of Natal Leadership Centre. Research in the field of organisational development, leadership and systems thinking remains a focal area.

Dr John E. Tookey graduated from the University of Bradford in July 1993 with a 2:1 degree in Technology and Management Science. He was subsequently invited back to the University of Bradford as a teaching assistant and to conduct research leading to a higher degree, graduating with a Ph.D. in Industrial Engineering in July 1998. The main topic of research during this period was in the implementation of Concurrent Engineering in aerospace supply chains. At the beginning of 1998, John moved to the Department of Building and Surveying at Glasgow Caledonian University as a research fellow to conduct research in construction procurement.

Derek Walker was born in Wales, UK in 1948 in the small seaside holiday resort town Porthcawl, jewel of the 'Welsh Riviere'. He studied at Bridgend Boys Grammar School, graduating in 1970 from the University of Glamorgan (at that time – Glamorgan Polytechnic) in Pontypridd, Cardiff. In January 1973 he left the UK to try his luck in Canada working as a quantity surveyor in London,Π Ontario.

In June 1978 he left Canada to study full-time for a Master of Science degree in construction management and economics at the University of Aston in Birmingham, UK. After finishing his studies he went to Australia and has lived in Melbourne since the end of 1978.

Derek joined Victoria University of Technology in January 1987 teaching construction and project management. In 1988 he joined RMIT first as course leader on the Bachelor of Applied Science in Construction Management and from 1991 until 1998 he was the head of department. He completed a Ph.D. which was awarded in 1994.

Foreword

It is with great pleasure that I introduce this best practice publication. The Working Commission 92 – Procurement Systems – is one of CIB's most dynamic groups which has gone from strength to strength over the past decade and this is a key deliverable to have emanated from this commission.

The topic of procurement systems is exceedingly important to the development of the global construction industry and it is the mechanism through which innovation – economic, technological, managerial – can be enabled and facilitated within the industry. The timing of this publication is particularly apposite with the current debate in Europe being stimulated by the publication of the Latham and Egan reports, which were critical of the construction industry's procurement practices, and with the massive infrastructural demands of China, Asia, Africa and South America needing to be informed of current best practice.

The publication is the most comprehensive of its kind in the field of procurement systems, dealing as it does with a whole range of issues, from client briefing to choice of contract strategy, from the impact of IT to the enhancement of organisational learning. This shows the strength of the links between W92 and other CIB working commissions and task groups and the bibliography is the most extensive there is in the field. This publication has been put together by CIB's leading researchers and writers, all of whom have been regular contributors to the W92 symposia. As a consequence, this volume represents the current body of knowledge in procurement systems and should be of interest to researcher and practitioner alike; the work is both academic and practical and is intended to fulfil both an educational and practical consultancy role.

One of the publication's strengths is its international nature. Contributions have been made from authors from as far apart as South Africa and the United States; Australia and the United Kingdom; and all of this coordinated from Hong Kong. This international perspective on best practice is one of the key strengths of this publication and of the W92 working commission. This is obvious in the range of topics dealt with in this volume; a whole range of key issues which are integrated into a coherent whole. For example, the links between IT, culture and organisational learning are pervasive within industry and these issues are clearly identified and integrated into procurement best practice. Uniquely, the key issue of developmentally oriented and appropriate procurement systems is addressed in this best practice guide.

Dr Wim Bakens
Secretary General
CIB

Preface

The purpose of this book is to gain access on behalf of clients, industry practitioners and researchers to the wealth of knowledge which has been built-up through nine years of activity of CIB Working Commission 92 – Procurement Systems. In 1989 laudable aims and objectives were established for the commission including:

'to research into the social, economic and legal aspects of contractual arrangements that are deployed in the procurement of construction projects;

to establish the practical aims and objectives of contractual arrangements within the context of procurement;

to report on and to evaluate areas of commonality and difference;

to formulate recommendations for the selection and effective implementation of project procurement systems;

to recommend standard conventions.'

The work of the commission has evolved from these initial aims and at the Symposium in Montreal in 1997 it was decided that the commission had developed a sufficient body of knowledge to produce a best practice guide to procurement systems. This would identify the key concepts and issues in this complex subject area. This has been achieved through a review of the proceedings of W92 and the commissioning of chapters by specialists in a range of areas that are central to procurement systems research and practice. The review, which forms the opening chapter of this book, summarises the major contributions to the proceedings and identifies significant themes. From these themes topics were identified which are of current interest and authors identified to produce chapters in those areas.

A brief look at the history of W92 indicates that it was recognised from the outset that some of the procurement systems subject matter had been included in the broader remits of other Working Groups, in particular W65 (Organisation and Management of Construction) and W55 (Building Economics). However, it was argued that there was a need for a working group that would serve the needs of the international research community in times of great change. In addition to long-standing concerns of international comparability and standardisation of contracts and contract procedures, the 1980s had brought significant changes in the legal, economic and social structures of states in both Developing and Developed Countries. Privatisation, in its many guises, had been effected not only in Europe and North America, but also in eastern Europe (through the transitions from socialist to capitalist systems), and in Africa and Asia (through Structural Adjustment Programs).

Some of the work which has been included in the Proceedings has been the primary output of projects, others are brought to CIB to increase international dissemination, while some are brought from areas defined by the CIB umbrella, others have clearly been promulgated within CIB. Much of the work is complementary, not only to that of W55 and W65, but also to that of TG 15 (Construction: Conflict Management and Dispute Resolution) and W82 (Future Studies in Construction).

Listed below are all the publications stemming from W92 proceedings (and their venues) which are referred to in this book. Some of the publications are still available and can be accessed from the CIB homepage, *http://cibworld.nl/*:

Anon (1991), Procurement Systems Symposium. Proceedings of Symposium held at Las Palmas, Gran Canaria, Spain, CIB Publication No. 145.

Bosanac, B. (ed.) (1990), Proceedings of Symposium held at Gradevinski Institut, Zagreb, Yugoslavia.

Cheetham, D., Carter, D., Lewis, T. and Jaggar, D.M. (eds) (1989), *Contractual Procedures for Building*, Proceedings of the International Workshop on 6–7 April held at the University of Liverpool, UK.

Davidson, C.H., and Tarek, A.M.(eds) (1997), *Procurement – A Key to Innovation*, Montreal, Proceedings of Symposium held at the Universite de Montreal, Montreal, Canada, IF Research Corporation.

Rowlinson, S. (ed.) (1994), *East Meets West*, Proceedings of Symposium held at The University of Hong Kong, Hong Kong, CIB Publication No. 175.

Taylor, R. (ed.) (1996), *North Meets South*, Proceedings of Symposium held at University of Natal, Durban, Republic of South Africa.

The book is divided into four main sections, the first of which sets the scene by way of introduction to the background of the work of W92, the issues which have emerged and a definition of key issues. Chapter 1 discusses strategic issues prevalent in the work of the commission: competition, privatisation, culture, trust, institutions, procurement strategies, contractual arrangements and forms of contract. Themes that have begun to surface in recent years are identified in the chapter as organisational learning, culture, developmentally orientated procurement systems, and sustainable procurement. Chapter 2 goes on to define procurement systems and discuss the nature of the construction process, the client, contract strategies and the application of management theory to these.

The second section deals with organisational issues in procurement systems. Green and Lenard, Chapter 3, discuss client organisation and the importance of strategic briefing. They then go on to identiy the ways in which value management differs from the traditional cost emphasis offered by value engineering. Two new methodologies of Group Decision Support are outlined. They argue that construction procurement cannot be understood in isolation and if progress is to be made it is necessary to view procurement as a constituent part of an integrated process that cuts across the established disciplines of the construction industry. The chapter concludes with a brief discussion of the need for innovative approaches such as partnering to enable organisational learning. Langford and his colleagues deal with organisational design in Chapter 4 and review past work in the area. They identify environmental influences on organisational design and identify communications and decision making issues during the construction phase. They discuss Jennings & Kenley's argument for greater explicit recognition of the non-technical determinants of project organisations and the advocation of the use of systematic thinking in designing and managing project organisations concluding with a discussion of organisational design and project success factors. Chapter 5 deals with the important issue of organisational learning in which the two Walkers explore the learning potential which is often not realised in the construction procurement process. They discuss, *inter alia*, innovation and organisational learning, sustainable construction principles and post project evaluation as an organisational learning tool.

The third section deals in some detail with those emergent issues in procurement systems; those issues which will have the greatest impact in changing how we view procurement systems. Fellows & Liu deal with the issue of culture and its many manifestations. They provide a theoretical perspective of how culture is perceived and discuss its potential impact on procurement systems and projects. Fellows is the co-ordinator for the CIB task group dealing with culture. Chapter 7 has been written by a number of authors and focuses on the important issues arising in developmentally oriented procurement systems, drawing on examples from South Africa by way of illustration. Issues faced by economies which are advancing are identified and the case is argued that current best practice in developed economies does not fit this different context. Chapter 8, by Walker and Rowlinson, investigates the impact that the Internet and IT in general has had on the construction industry. Results of two surveys of IT usage are reported and the authors postulate the impacts that these are likely to have on the future development of procurement systems.

The fourth and final section deals with procurement systems in practice, addressing such issues as evaluation of contractors and selection of appropriate procurement systems. In Chapter 9 Miller deals with the issues which need to be addressed in procuring portfolios of projects for infrastructure provision. Miller indicates that knowledge, experience, and judgment are required for success, and that choice of project delivery method affects the timing, nature, and scope of innovation in technology, design, construction, operation, and finance. He discusses the relationship between project delibery method and financing vehicle. Chapter 10, written by Kumaraswamy and Walker, deals with performance criteria that are used to evaluate construction contractors. They draw on extensive research to provide the basis for a methodology to be used in the evaluation process. The topic of partnering is dealt with by Matthews in Chapter 11 and the principles in producing a partnering agreement are laid out in detail, based on research undertaken by the author in the United Kingdom. The section, and book, concludes with a chapter by Rowlinson on the subject of selection criteria for procurement systems. Rowlinson reviews previous work aimed at producing a set of selection criteria, both manual and IT assisted, and concludes the chapter with a framework within which a selection criteria methodology can be produced to suit particular situations.

This book has been so designed to appeal to researcher, student and practitioner alike. However, it is a rather comprehensive piece of work and for those looking for a rapid understanding of the area then practitioners might wish to go to Chapters 2, 9, 10, 11 and 12 first, students to Chapters 1 and 2 and researchers to Chapters 3 to 8 initially. However, readers are advised to take the book as a whole as it covers the whole range of procurement issues.

It would be impossible to produce a book such as this without the active participation of a host of people. We would like to thank the following people for their contributions to this work and apologise to any that we neglect to mention. First of all, we would like to thank all the contributors who have put in an enormous amount of work to ensure that each individual chapter in this book is at the forefront of research and practice in procurement systems. In addition, we would like to thank a number of people who have been heavily involved in the production of this book. First and foremost we would thank Kannex Chu for all the many hours of hard work she has put in typing, correcting, formatting and printing the proofs of this text. We would like to thank Jason Matthews, Florence Phua and 'Hari' Hadikusumo Bonaventura of the University of Hong Kong for the enormous amount of work they put in proof reading, checking citations, researching specialist issues and producing various drawings, diagrams and tables for the text. It would be remiss of us to forget to mention all of those who actively participated in W92 symposia; it is their work and ideas which have helped form the body of knowledge on which we have all drawn in making our contributions to this book. We should especially thank the organisers of the symposia over the past nine years and the editorial committees that have done such a fine job in refereeing papers and helping to raise the standard of work presented at W92 symposia to the current high level. Additionally, the support of the CIB Secretary General, Dr. Wim Bakens is gratefully acknowledged and last, but not least, the support of the founding coordinator, David Jaggar, must be mentioned. Without Professor Jaggar's energy and enthusiasm W92 would not have developed to the extent that it has today. Finally, we would like to acknowledge the contribution made by our students, mainly at Masters and Ph.D. level but also at undergraduate level, in their dissertations and assignments. They are too numerous to mention individually but have studied with us at, *inter alia*, the Universities of Hong Kong, Bath, Brunel, Cape Town, Liverpool John Moores, Salford, Reading, UTS Sydney, RMIT Melbourne, MIT and elsewhere.

Abbreviations

ACII	Australian Construction Industry Institute
ADOT	Arizona Department of Transport
ADR	alternative dispute resolution
AEC	architectural, engineering and construction (enterprises)
AGC	Associated General Contractors (of America)
AIC	Australian Industry Commission
ANOVA	analysis of variance
BAA	British Airport Authority
BIFSA	Building Industries Federation of South Africa
BOO	build–own–operate
BOOT	build–own–operate–transfer
BOT	build–operate–transfer
BP	British Petroleum
BPO	behaviour–performance–outcome
BPR	business process re-engineering
BREEAM	Building Research Establishment environmental assessment (method)
CAD	computer-aided design
CAE	computer-aided engineering
CIB	International Council for Building Research Studies and Documentation
CIDA	(Australian) Construction Industry Development Agency
CII	Construction Industry Institute
CIRIA	Construction Industry Research and Information Association
CM	construction manager or construction management
CONQUAS	construction quality assessment system
COPS	conventionally orientated procurement system
CRT	Construction Round Table
CSF	critical success factor
CSIR	(South African) Council for Scientific and Industrial Research
CTP	construction time performance

DB	design–build
DBB	design–bid–build
DBFO	design–build–finance–operate
DBO	design–build–operate
DBOM	design–build–operate–maintain
DBOT	design–build–operate–transfer
DCF	discounted cash-flow (model)
DOPS	developmentally orientated procurement system
E&CC	Engineering and Construction Contract
EC	European Communities
EDI	electronic data interchange
EPC	engineering–procurement–construction (sector)
EMS	environmental management system
FT	fast track
GATT	General Agreement on Tariffs and Trade
GDS	group decision support
HKGSAR	Hong Kong Government Special Administrative Region
HTCPS	hybrids of the traditional construction procurement system
IAI	International Alliance for Interoperability
ICE	Institution of Civil Engineers
IDM	integrated data model
IRP	issue-resolution policy
IT	information technology
JCT	Joint Contracts Tribunal
LRA	linear responsibility analysis
MC	management contractor
NEC	New Engineering Contract
NEDO	National Economic Development Office
NGO	non-governmental organisation
NIC	newly industrialising country
O&M	operate and maintain
OHS	occupational health and safety
OOCAD	object-orient computer-aided design
PASS	performance assessment and scoring system
PP	parallel prime
PVQ	personal values questionnaire
QA	quality assurance
QS	quantity surveyor
RCF	Reading Construction Forum
SAVE	Society of American Value Engineers (SAVE International)
SC	subcontractor
SHE	safety, health and environment
SSM	soft systems methodology
TCP/IP	transmission control protocol/Internet protocol
TCPS	traditional construction procurement system

TKY	turnkey
TQM	total quality management
TMO	temporary multi-organisation
UE	uncertainties pertaining to the working environment
UN	United Nations
UR	uncertainties pertaining to related decision fields
USAC	United States Army Corps
UV	uncertainties pertaining to guiding values
VE	value engineering
VM	value management
VR	virtual reality
WTO	World Trade Organisation

Part I

Introduction to procurement systems

1 Strategic and emergent issues in construction procurement

Peter McDermott

Introduction

The purpose of this chapter is to explore and explain the strategic issues of construction procurement which have arisen through the published work of the Commission CIB W92 – Procurement Systems of the International Council for Building Research Studies and Documentation. The contributions have been made from very different theoretical and cultural perspectives; the contributors have come from all continents and from different economic, legal and political systems. The significant strategic issues prevalent in the work of the Commission have been identified as competition, privatisation, development, culture, trust, institutions, procurement strategies, contractual arrangements and forms of contract. Emergent themes – those which have begun to emerge in recent years and which are expected to grow significantly – have been identified as organisational learning, culture, developmentally orientated procurement systems and sustainable procurement.

This chapter is a development of work reported elsewhere (McDermott and Jaggar 1997; McDermott *et al*, 1997). The Commission was founded in 1990 as a result of an international workshop held at the University of Liverpool in 1989 entitled International Workshop on Contractual Procedures. Since then, in addition to Commission meetings, there have been six separate Symposia, in 1989, 1990, 1991, 1994, 1996 and 1997. In 1993 and 1997 the Commission operated in association with W65 (Construction Management). Much of the work of the commission is complementary, not only to that of W55 (Building Economics) and W65, but also to that of TG15 (Construction: Conflict Management and Dispute Resolution), W82 (Future Studies in Construction) and TG23 (Culture and Construction).

What is procurement?

The procurement concept in construction has been defined in many ways. Hibberd (1991) began with the general definition of the term procurement offered by the *Oxford English Dictionary*, 'the act of obtaining by care or effort, acquiring or bringing about', and then argued that the concept of procurement can raise awareness of the issues involved both in challenging generally accepted practices and in establishing strategies.

Others, including Mohsini and Davidson (1989: 86), have attempted more sophisticated definitions – 'the acquisition of new buildings, or space within buildings, either by directly buying, renting or leasing from the open market, or by designing and building the facility to meet a specific need'. At the Montreal Symposium in 1997 the Commission was challenged to debate hypotheses which attempted to establish the relationship between procurement and innovation (Davidson, 1998). For the purposes of this debate, the following definition was accepted: 'Procurement is a strategy to satisfy client's development and/or operational needs with respect to the provision of constructed facilities for a discrete life-cycle' (Lenard and Mohsini, 1998: 79). This sought to emphasise that the procurement strategy must cover all of the processes in which the client has an interest, perhaps the whole lifespan of the building.

However, it has been argued that for some research purposes the usefulness of definitions such as this is limited (McDermott and Jaggar 1991). For example, as a means of comparing projects or project performance across national boundaries it is limited to developed market economies. This conclusion is supported by Sharif and Morledge (1994) who have drawn attention to the inadequacy of the common classification criteria for procurement systems (that is, traditional, management, design and build, etc.) in enabling useful global comparisons.

Even comparisons between developed economies are fraught with difficulties. Latham (1994: 5), in a review of procurement and contractual arrangements in the UK, has noted the difficulty of drawing conclusions from existing studies, stating that 'some international comparisons reflect differences of culture or of domestic legislative structures which cannot easily be transplanted to the UK'. Davenport (1994) has reported that the French do not recognise the British and North American concept of procurement.

It has been argued that two of the key assumptions contained within the common definitions – those of client choice and the availability of a range of procurement options – render the procurement concept, as defined, irrelevant to Third World countries (McDermott *et al*, 1994). However, a working definition of procurement was developed by CIB W92 at its meeting in 1991, defining it as 'the framework within which construction is brought about, acquired or obtained' (unpublished document). This definition served a useful purpose as it is both broad, encouraging a strategic interpretation, and neutral, being applicable not only to developed, market economies.

Theoretical foundations

The formal aims of CIB W92 (see the Preface) clearly suggest that procurement is a social science and implies that the disciplines of history, sociology, economics, psychology, law and politics can all make a contribution to fur-thering understanding. Rarely, however, do researchers or practitioners make their approach explicit. Green (1994) argues that research has reflected the positivism of functionalist sociology and has largely ignored the validity of naturalistic inquiry. This means that the research method-ologies adopted tend towards the establishment of causal relations from a distance (having assumed that there is an objective reality which exists in-dependently of human perception) rather than to the participation of the researcher in what is being researched.

Green attempts to identify and characterise shifts in the paradigms of procurement practitioners during the 1980s and early 1990s in the UK. Although very few practitioners give any explicit consideration to their operative paradigm, Green maintains that a 'default paradigm of practice' can be identified from behaviour in the field. Green draws upon the work of Morgan (1985) who identified eight different organisational metaphors, amongst which practitioners might recognise their own views of the world. These metaphors portray organisations as machines (which also relates to scientific management), biological organisms (which also relates to the systems approach), brain (or experiential learning), competing cultures, political systems, psychic prisons, states of flux and transformation, and instruments of domination.

Much of the work of practitioners and researchers in CIB W92 can be identified within the machine or biological metaphors. For example, the systems and contingency approaches are used extensively (both explicitly and implicitly). Hughes (1990), drawing upon a tradition of using the sys-tems approach to analysing the organisation of construction projects, starts with the premise that buildings are procured through organisational sys-tems. He describes various means whereby flexible procurement systems can be designed which are appropriate to each project. Naoum and co-workers (Naoum 1990; Naoum and Coles 1991; Naoum and Mustapha 1994) used systems-based models in comparing the performance of alterna-tive procurement systems. Also, Carter (1990) outlines the use of data-flow diagrams and activity profiles as a means of improving the management information systems of designer and contractor organisations.

Many researchers have conducted their work within a socio-technical framework. Jaggar and I (McDermott and Jaggar 1991; McDermott 1996; Newcombe 1994) and Jennings and Kenley (1996) have emphasised the relevance of the 1960s work of the Tavistock Institute for procurement research now. Jennings and Kenley (1996), for example, have argued that the dominant functionalist paradigms of the 1980s failed to extend beyond the technical logistics of procurement and into a consideration of the social aspects of organising for procurement.

The contingency approach is evident in the work of Swanston (1989) and Singh (1990), amongst others. Many researchers draw upon the models and insights provided by the transaction-cost or markets-and-hierarchies approach developed by Williamson (1975). Doree (1991) notes the drift towards the 'contracting-out' of design services by public clients and asks the question: When should unified governance structures (in-house production) and when should market governance structures (contracting-out) be applied? He concludes that although the move towards contracting-out can be justified on the basis of design production efficiencies, the equation is not complete if the transaction costs of operating in the market and the opportunity costs (of, for example, increased life-cycle costs) are not considered.

Chau and Walker (1994) investigated the nature of subcontracting in the Hong Kong construction industry and concluded, *inter alia*, that the decision to subcontract is not random but is predicated on the attempt to minimise transaction costs. Alsagoff and I adopted a similar methodological approach in investigating the true level of infiltration of relational contracting in the UK construction industry (Alsagoff and McDermott 1994). Cheung (1997) uses a transactional analysis to construct a model for determining the most appropriate form of dispute resolution procedure.

The business process re-engineering (BPR) paradigm and the so-called new (lean) production philosophies which have penetrated other industries have been investigated in construction. For example, Baxendale *et al.* (1996) investigate the implications of concurrent engineering for roles and relationships within procurement systems, and Lahdenpera (1996) argues that procurement needs to be re-engineered in a fundamental way rather than to continue to study incremental and non-fundamental issues. Green (1997) argues for a soft systems interpretation of BPR for application in construction. BPR, he suggests, as currently practised, is relevant only where the problems are easily identifiable. In complex circumstances, as in construction, these techniques will not lead anywhere. Egbu *et al.*(1996) sought to compare the procurement of project work in ship refurbishment with that in the construction industry. BPR (Mohamed 1997) and lean production (Alarcon 1997) are both the subject of significant research activity elsewhere.

Jennings and Kenley (1996) also emphasise the importance of recognising the theoretical underpinnings of procurement research. They argue that procurement systems go beyond technical logistics and that it is the perception and response to project objectives by organisations which is a key determinant of procurement system suitability. Liu (1994) takes this further and discusses a cognitive model of the procurement system where the goal–performance relationship is paramount. Liu uses conjoint analysis as a means of illustrating this.

Kumaraswamy (1994a) discusses the appropriateness of developed countries' procurement systems when applied to less-developed countries and argues that a sustainable and synergistic procurement strategy must be developed in such situations. The power paradigm (Newcombe 1994: 245) suggests that selection criteria for procurement systems are less important than the realisation that 'procurement paths create power structures which dramatically affect the ultimate success of the project'. Using this paradigm Newcombe criticises the fragmentation and friction evident in the traditional system. This is further supported by Walker (1994) in that he draws the conclusion that project construction speed is strongly determined by how well clients relate to the project team.

The importance of conceptualising the procurement problem is that lessons can be learnt in one context and transferred to another. Although practitioners clearly need to focus on solutions the role of the research community is to provide the conceptualisation and theory which will lead to the development of best practice and to satisfying the technology transfer objectives of the CIB.

Development and privatisation

During the 1980s the technical and academic press reflected client concerns of project performance in construction, with much debate concerning international comparability and standardisation of contracts and contract procedures. This was in an environment of significant changes in the legal, economic and social structures of states in developing and developed countries. Privatisation, in its many guises, had been effected not only in Europe and North America but also in Eastern Europe (through the transitions from socialist to capitalist systems) and in Africa and Asia (through structural adjustment programmes).

Jackson and Price (1994) have detailed how the process of market liberalisation has conflicted with developmental goals worldwide. The conflict is evidenced in practice, policy and ideology. Within the field of construction performance the emphasis has been placed on practice and policy, not on ideology. Examples from W92 proceedings range from the impact of compulsory competitive tendering for design professionals in the UK to the effect of structural adjustment programmes in Africa. The concern with broader developmental goals means that procurement systems which consider more than speed, quality, price competition and certainty, and risk transfer are needed.

Procurement systems must be appropriate to circumstances. Kumaraswamy (1994a) drawing upon examples from Sri Lanka, argued that the achievement of 'technology transfer' (which itself is often misconceptualised) from developed to developing countries has been thwarted by the use of inappropriate procurement systems. For example, transplanted prequalification systems, unsustainable capital inputs and inflexible packaging of contracts within major projects resulted in the marginalisation of the local industry and hence in distortions to the construction industries of developing countries.

The paradigm shift argued for by Kumaraswamy (1994b) is supported by Taylor and Norval (1994) who present a framework for the establishment of procurement systems which build-in industry developmental goals in South Africa. No longer can South Africa rely solely upon the paradigms of the developed world. It must develop procurement systems which consider more than speed, quality, price competition and certainty, and risk transfer. The procurement systems must encourage, *inter alia*, appropriate, people-intensive technology and processes, learning and skill development. In these circumstances, the process of procurement assumes a greater status than it is normally afforded. The process-related goals become as important as the product-related goals (Martins and Taylor 1996).

Within the so-called developed world much of the attention is focused on narrower procurement and contractual relationships. Lahdenpera (1994) describes the response of the vested interest groupings to a Finnish initiative aimed at introducing a more appropriate generic structure to the overall building process organisation.

Current debate in the UK concerns the extent to which the Latham Report (a joint government–industry report) implemented by the Construction Industry Board, is succeeding in changing the adversarial culture of construction. Perhaps, the Latham Report process should be seen more as a means of coping with the earlier cultural shift of market liberalisation than as a driver for cultural change itself. The traditional procurement and contractual procedures which have been swept aside by market liberalisation have needed to be replaced with other control mechanisms. Weak forms of control, such as reliance upon craftsmanship and professionalism, had been replaced with strong forms of control, such as contract enforcement and litigation. The Latham Report endorsed initiatives such as alternative dispute resolution (ADR) and partnering as a means to replace the culture of contentiousness and to rebuild the trust which had been swept away.

The narrower aims of the Latham Report in the UK have prompted further detailed analysis of the procurement and contractual arrangements which have fed directly into W92. For example, Potts and Weston (1996) and Potts and Toomey (1994) discuss the application of risk management techniques and alternative payment systems, respectively, both in sympathy with Latham Report recommendations.

Market liberalisation and privatisation policies have not only been effected in developed countries such as the UK. Suite (1996) described the procurement responses of the Government of Trinidad and Tobago to the calls by the multilateral and bilateral aid agencies for privatisation and divestment as a precaution for further loans. The problem faced by many governments – how to commission projects without having to make any capital payments during the construction and maintenance periods of each project – had already been faced by the design of a project financing package on behalf of the Government by a consortium of private banks. Initiatives such as this and the Private Finance Initiative in the UK (Akintoye and Taylor 1997), by which the private sector finances public sector projects, change the nature of the procurement process from that of the purchase of a capital asset to the purchase of a service. The procurement strategy which meets these needs is the development of design–build–finance–operate (DBFO), build–own–operate (BOO) and 'turnkey' projects. These alternatives are described in more detail in the section on classification and choice in procurement systems, but first the role of culture will be discussed.

The role of culture: trust and institutions

A review of the work of W92 exposes the extent to which culture (to which the two dimensions of trust and institutions are attributed) influenced procurement. This was evidenced regardless of the economic or political system. This section briefly explores the meaning of trust and institutions (including government, intermediate agencies and international agencies).

Fukuyama (1995) has argued that the ability of nations to compete is conditioned by one pervasive cultural characteristic – the level of trust in society. Trust was seen by Latham (1994) as the gatekeeper to any real progress in improving procurement and contractual relations in the UK construction industry.

Culture itself can be distinguished conceptually from social structure (although in practice the two are almost inseparable): 'Culture in this sense is restricted to meanings, symbols, values . . . religion, ideology . . . [whereas] social structure . . . concerns concrete social organisations such as the family, clan, legal system, or nation' (Fukuyama 1995: 34). The capability of communities to form new associations and to cooperate Fukuyama calls spontaneous sociability.

The potential for spontaneous sociability is dependent upon a network of social and political institutions (or social structure). Hutton (1994: 20), in identifying areas of UK weakness, argued that 'the degree to which an economy's instutions succeed in underpinning trust and continuity is the extent to which long term competitive strength can be sustained'. A major thrust of the Latham Report (Latham 1994) in the UK has been the attempt to rebuild trust in the construction industry. This has been attempted both through the advocation of partnering at the project level and through encouraging the restructuring and realigning of the existing client, contractor, subcontractor, supplier and consultant institutions.

The role of intermediate institutions in establishing procurement policy and practice is alluded to by members both from developed and from developing countries. Ng (1994) presents two cases concerning joint ventures between local government agencies and non-governmental organisations (NGOs) in Guangxi in the People's Republic of China. His conclusion, that people participation which enables community development is essential in the procurement process, concurs with that of Martins and Taylor (1996).

How the establishment of construction industry development boards affect procurement policy and practice are alluded to by Walker (1996), who writes from an Australian perspective, and by Allin (1990), and Ofori and Pin (1996), each based in Singapore. Ofori and Pin summarise the advantages of such intermediate institutions as encouraging the public sector to divide its projects into suitable sizes (to allow participation by small, local contractors), to use locally developed materials and to formulate contract procedures and documents which are easy to use by local industry.

Others report on the direct involvement of governments in determining procurement policy and practice. Aziz and Ofori (1996) show how the Malaysian government, by tying development objectives to its privatisation programmes, has been able to stimulate the growth of Malaysian national contractors, who have subsequently captured a significant market share in neighbouring countries.

Mustapha *et al.* (1994) show how the liberalisation of the economy in Turkey, which was implemented through the promotion of design–build–own–operate–transfer procurement systems, ultimately resulted in increased activity by Turkish contractors in neighbouring countries. By contrast, Mukalula (1996) reported on the effects of a structural adjustment programme on procurement policies and practices in sub-Saharan Africa. With particular reference to Zambia he shows how the private sector failed to invest at sufficient levels in the formerly state-owned enterprises.

Torkornoo (1991) details the incentives which are offered to overseas investors in Ghana, where the construction industry is designated a 'priority area'. The incentives are mainly in the form of import duty waivers on capital and of income tax exemptions. Although there are quotas for expatriate personnel, there are no controls over the choice of contract procedures or forms, which remain private decisions.

The transition of the former USSR to a market economy, and to market relations in the construction sector, is detailed by Klimov *et al.* (1990). Problems of completion on time and to budget had frequently arisen as, under the centrally planned system, the annual distribution of investments for construction did not always materialise and sometimes contractors were allocated too much work to complete given their existing resources. Klimov *et al.* detail the moves to privatise construction companies and to introduce antimonopoly laws. These are seen as prerequisites for the introduction of new legal and institutional frameworks for investment, new principles of pricing projects and new standard contracts. Further details on how transformation of state-owned companies into joint stock companies can be achieved is illustrated through a case analysis from Yugoslavia (Vukovic and Marinic 1994).

Craig (1996) reviews the public procurement regulations of the European Communities (EC), the World Trade Organisation (WTO), the World Bank and the United Nations (UN). Whilst all of the regulations are designed to promote competitive tendering and to prevent the use of technical standards as non-tariff barriers, only WTO and the General Agreement on Tariffs and Trade (GATT) encourages positive discrimination on behalf of developing countries.

Lloyd-Schut (1991) outlines the legislation introduced by the EC, in its moves to create a Single European Market. Member states are required to bring into domestic law the requirements of directives made by the EC. Lloyd-Schut, *inter alia*, shows how the EC aimed to establish common rules for all public sector purchasers (from local authorities to government departments, and including certain private sector companies who hold 'special or exclusive rights' with a member state) in the advertising and awarding of contracts for certain projects. Purchasers were given the right to reject a tender, exercising 'Community preference', if the tender had less than 50 per cent EC content.

Hodgson (1994) discusses the impact of EC Directive 83/189, which lays down a procedure for the provision of information in the field of technical standards and regulations. The response of the Department of Transport in the UK resulted in a significant change in its approach to procurement. Its standard specifications and standard contracts shifted the discretion on choice of materials and products away from the engineer to the contractor.

Lavers (1991: chapter 7, page 9) argues that legal mechanisms alone are incapable of ensuring the achievement of quality as the directives and regulations are predicated upon the assumption that in developed, industrialised states (such as the EC Member states), with construction industries which are in the main highly sophisticated, 'there will be both the capacity and will to achieve a high degree of compliance with the standards produced by these procedures'. In other words, it is the social system which underpins legislation and makes it effective. European mechanisms would be inappropriate in developing countries.

Procurement systems: classification and choice

Hibberd (1991) has argued that no standard definitions and classification of procurement approaches have become generally acceptable, quite simply because there are no formal structures or agreement on the terms. Further, as this is so, it also follows that the term 'procurement system' implies a degree of scientific rigour which does not exist. As such, he argues, either the term 'procurement path' or 'procurement approach' would be preferable.

Hibberd (1991) found that UK practitioners identified with the following eight procurement paths despite the fact that some of them share a significant number of characteristics: conventional or traditional, management contracting, design and build, two-stage conventional, construction management, British Property Federation System, prime cost, and develop and construct.

The categorisation of procurement paths by the National Economic Development Office (NEDO 1983) into traditional, design and build, design and manage, management contracting and construction management has found favour with many UK-based researchers. For example, Hamilton (1990) reviewed the use of these procurement systems in the UK. Others have reported research which has resulted in the construction of decision-making tools for clients [see, for example, Singh (1990) on the USA, and Swanston (1989), on the UK].

The difficulties to which Hibberd (1991: chapter 8, page 3) refers, of understanding the inherent virtues and characteristics of procurement paths, however defined, might arise from what Ireland (1984) referred to as 'virtually meaningless distinctions between nominally different procurement methods'. In other words, in all of the attempts to distinguish between procurement systems, it is forgotten that they are actually more similar than different! Dulaimi and Dalziel (1994) report on work elsewhere which identified 59 hybrids of design and build alone.

Gow and Fenn (1989) point out that what is now perceived as the 'traditional' approach to procurement in the UK only became so after the Industrial Revolution, before which 'separate trade contracting', akin to modern construction management, was the norm. Indeed, separate trade contracting was extensive in Scotland until the 1960s, after which English methods came to the fore.

Modern 'construction management' (where the client appoints a construction manager on a fee basis and enters into separate, direct contracts with trade contractors) developed in the USA and found markets in Europe and Australia. Lam and Chan (1994) conclude that in Asia, where construction management is still a relative newcomer, it is likely to increase in use, but only in the more developed countries.

Ogunlana and Malmgren (1996) concluded that the use of construction management in Thailand is likely to grow. This was in spite of experiences of delays in projects which have used the system since its introduction 10 years ago.

Naoum and Coles (1991; Naoum 1990) sought to establish whether the choice of procurement path influenced the performance of the project. Focusing on the management contracting path (which is similar to the construction management path except the management contractor enters into direct contractual relationships with the trade contractors) and the traditional path, they concluded that the former system performed better when time was of the essence and when the project was highly complex.

Wide dissatisfaction with traditional approaches to procurement (especially in terms of cost and time) together with acrimonious conflicts between the various parties involved have fuelled the rapid expansion in use of design and build and its variants [see, for example, Saito (1994) on Japan; Dulaimi and Dalziel (1994) on the UK; Smith and Wilkins (1994) on the UK and the USA; Dreger (1993) on the USA; Gunning and

McDermott (1997) on Northern Ireland; and Hashim (1997) on Malaysia]. The benefits of design and build are often argued in terms of closer integration of design and construction teams. Dulaimi and Dalziel (1994) were able to demonstrate not only improved integration amongst project team members but also an increased level of synergy.

As referred to earlier, a further significant development has been the rapid increase in the use of build–own–operate–transfer (BOOT) arrangements (also known as the concession method). This concept, established more than a century ago to construct canals and railroads, was sought and encouraged by governments as a means of obtaining private sector finance for projects, such as infrastructure projects, which in modern times have been a drain on the finances of the public sector. The potential for private financing of projects has been attractive to governments in the First and Third World, in Eastern and Western Europe, in Asia, in the USA and in Africa. However, it has been especially so for the provision of infrastructure in developing nations [see, for example, Mustapha *et al.* (1994) on Turkey; Tam *et al.* (1994) on China, the Philippines and Thailand; and Ogunlana (1997) on Thailand].

Contractors became involved (normally as part of a consortium) to guarantee involvement in the construction work. In consideration for raising the finance for the project the consortium would be granted a concession to operate the facility for an agreed period of time, after which the function is required to be transferred back to the state. This reliance on a future stream of income as a reward to the investors led the BOOT method to be advocated mainly for schemes for which there was a clearly defined income source, for example a tolled road, bridge or tunnel.

However, this has changed in recent years, with the BOOT strategy being adopted for the construction of prisons, hotels, telecommunications systems, the processing of water and waste water and the use of coal, gas and oil for power generation. Merna and Smith (1991) argued that it is appropriate even where there is no direct revenue source, such as in public sector schools and hospitals and sheltered housing. In the UK it is now government policy to require all public sector projects to be market tested for private finance before the public purse is opened. Although the principle of this policy has been accepted enthusiastically by some in the industry, there has been almost unanimous criticism of the speed and effectiveness of its implementation.

Tiong (1990) has described a typical contractual network for a BOOT-style project. Normally at the centre will be a joint venture or project company legally constituted in the host country. The project company will need to establish contractual relationships in anticipation of the length of the concession, which would typically be 30 years for an infrastructure project. In addition to the concession agreement with the host government or agency, loan agreements with the banks, shareholder agreements with investors, offtake contracts with the users of the facility, operation agreements with the operators and construction contracts with the builders all need formulating.

As with all procurement systems, the key to a successful BOOT strategy is the identification and evaluation of the risks involved, followed by their allocation to the parties most able and willing to accept them. As more experience is obtained in the use of this strategy, potentially all parties, guided by their professional advisers, will be able to make more rational decisions regarding risk acceptance and apportionment. It is likely that standard agreements will be developed and accepted which will lower the costs and clarify the contractual establishment.

According to Tiong (1990) the prerequisites for success for sponsors (that is, the concession company) on concession projects in the international arena are strong government support, a stable currency and a stable economic system. Tiong identified those factors which potential sponsors should address specifically to ensure success in the winning of a concession. The factors – critical success factors – were:

- entrepreneurship;
- choice of the right project;
- a strong team of stakeholders and suppleness in their relationships;
- the use of imaginative technical solutions;
- the tendering of a competitive financial proposal;
- other special features.

Host countries, particularly those in the Third World, may procure an inappropriate project as a result of inequalities in the commercial arrangements (McDermott *et al.* 1994). Kumaraswamy (1994a) explains that whereas some projects were formerly financed by clients or 'promoters', creative financing strategies are now used, for example, on large-scale infrastructure projects. Contractors with underutilised resources may team up with the idle capital of banks and commercial interests to promote ventures of the concession or build–operate–transfer (BOT) type. The traditional roles and demarcations of the parties and their agents become more fuzzy as contractors take the lead and also employ the design consultants whereas the client commissions an 'independent checking engineer'. In addition, the potential for conflict of interest occurs where the appointed contractor has a share in the concession company. Although such BOT ventures have been relatively successful on tunnel projects in Hong Kong and some power stations in China and the Philippines, the need for developing frameworks and guidelines is suggested by project breakdowns, as, for example, in the Bangkok Expressway Project.

In the context of the developing countries Kumaraswamy (1994b) argues for a longer-term overview to be encouraged to promote domestic contractor and general economic development. For example, in Sri Lanka in the late 1970s donors and funding agencies for the Mahewli Irrigation and

Hydropower Scheme formulated packages that restricted most consultancy, contracting and supply services to foreign agencies.

There is now some degree of consistency in the classification of procurement systems which appears in publications. However, the choice of an appropriate procurement system is not always given to clients for the reasons detailed. Indeed, Bowen *et al.* (1997) found that clients who had successfully chosen an appropriate procurement system for their needs were more likely to have relied upon good luck than good advice.

Contractual arrangements and forms of contract

The original focus of the embryonic W92 was contractual arrangements and forms of contract. Although W92 has broadened its remit, these subjects are still central to its work. A substantial proportion of the proceedings are still devoted to subjects such as the decision to tender (Eastham 1990), contractor assessment and central contractor registration schemes (Allin 1990; Ramsay-Dawber 1996), tendering practice (Wahlstrom 1989), subcontracting arrangements (Cottrell 1989; Brochner 1989; Ndekugri 1989; Pasquire 1994) and the use of knowledge-based systems in contracts (Pollock and Rees 1989; Hibberd *et al.* 1994; McCarthy and Wang 1994).

The use of standard forms of contract has been reported extensively [see, for example, Klimov *et al.* (1990) on the former USSR; Dimkic (1990) on the former Yugoslavia; Robinson (1990) on Singapore; Brochner (1989) on Sweden and Denmark; Torkornoo (1991) on Ghana; Gidado (1996) on Nigeria; and Leung *et al.* (1996) on Hong Kong].

Fellows (1989: 17) hypothesised that 'the environment in which construction contracts are produced is outmoded, and is the cause of, rather than a solution to, many of the industry's recent and current difficulties'. Although the hypothesis was developed in the UK context, he argues that it has wider application because of the influence of British standard forms worldwide. The standard forms, produced both by the Joint Contracts Tribunal (JCT) and by the Institution of Civil Engineers (ICE), whether appropriate or not, have been used extensively (either amended or not) in environments for which they were not designed.

For example, the standard form of contract published by the Singapore Institute of Architects, although originally drafted by an English lawyer, is biased towards the client. Unlike the forms created by the JCT in the UK it does not benefit from supply-side representation (Robinson 1990).

Currently in the UK, in response to the Latham Report (1994), the JCT forms are being re-examined, and the Engineering and Construction Contract [previously known as the New Engineering Contract (NEC)] is being encouraged as an alternative. Williams (1991) has described the background to the development of the NEC. He drew attention to the dissatisfaction of many UK civil engineering clients with the existing standard contract forms. The ICE Conditions were being routinely amended to suit clients' own needs (for example, see Hill 1991) and some were turning to forms of contract designed for use in other industries. The NEC was introduced in an attempt to overcome some of these difficulties. More recently it has been adapted for use on building works.

However, in spite of the range of work submitted to the Commission, there is still plenty of research required in identifying international best practice and appropriate implementation procedures.

Procurement: culture and conflict

It is evident from the foregoing discussion that research in the area of procurement systems has moved from a hard, technical systems approach into a much more soft-systems-based set of paradigms which appear to have strong potential for explaining the differences in performance across projects. The concept of culture has become an important issue in analysing procurement systems (Rowlinson and Root 1997; Crook 1996).

Rowlinson and Root (1997) were surprised to find that the impact of conditions of contract on performance was very limited. The explanation that project prehistory and prior working relationships (and, to an extent, procurement form adopted) have the most significant impact on project culture is important. The view often has been expressed that the conditions of contract are only necessary when a dispute arises and that good working relationships can avoid this scenario. Thus, development of a positive project culture, even before a contract is let, is the best means of ensuring a smooth-running project. If each participant comes to the table at the first meeting with an agenda of pre-project grievances then conditions of contract will only mark out the battle lines; they will not change attitudes. The view put by one of the interviewees in the study, that paying a fair price for a good job is the only way to ensure a cooperative and proactive project team, is well made.

The relationship between procurement, construction claims, dispute management and arbitration is widely reported by, amongst others, Fenn (1989), Abdel-Meguid and Davidson (1996), Heath and Berry (1996) and Conlin *et al.* (1996). Fenn (1989) summarises the means of settling disputes on construction projects in the UK, mainland Europe and the USA. There are strong cultural differences between the litigation, arbitration, adjudication and conciliation procedures which have developed in these countries compared with those of Japan, where many of the typical Western conflicts do not arise.

Abdel-Meguid and Davidson (1996) tested the hypothesis that construction claims, as a manifestation of interorganisational disputes, are the result of a poor choice of procurement strategy. A tentative correlation was identified between the traditional procurement strategy and delay claims, based on the variables 'percentage time overrun' and 'percentage cost over-run'. Heath and Berry (1996) report on a research project in progress which has similar objectives. Starting from the premise that there is a correlation between claims and procurement strategy, they trace the origins of conflict through the various stages of construction projects. The aim is to modify procurement procedures to avoid conflict.

Conlin *et al.* (1996) are also tracing the causes of conflict through the duration of projects. Their tentative conclusions are, *inter alia*, that certain conflict groups (namely, payment and budget, performance, delay and time, negligence, quality and administration) can be associated with particular procurement systems.

Liu and Fellows (1996) have explored the relationship between the procurement process and mechanisms and the mindset, values, beliefs and behaviours of the people involved. They are seeking to raise the awareness of culture amongst practitioners as a means of encouraging participation in the setting and achieving of project goals.

Martins and Taylor (1996) examined the concept of culture in the context of the procurement policies of the Re-construction and Development Programme in South Africa. Their argument, that procurement and contractual practices should be based on principles appropriate to indigenous cultural values, was rooted in an anthropological methodology. A brief review of past procurement policies concluded that 'the technocratizing of procurement has involved a massive de-legitimizing of alternative knowledge systems rooted in the traditions of local communities and a disenfranchising of these communities' (unpublished paper). Edwards and Bowen (1996) describe the situation in South Africa and point out that communication is a key element in determining success in a newly developing economy. This reflects the emphasis on culture which is well explained in the work of Liu and Fellows (1996).

Environmental sustainability and procurement

Following an initial emphasis on contractual and procedural issues, organisational and managerial considerations are now firmly embedded in the mindset of researchers and, to a large extent, practitioners. However, there has been little work within W92 concerning the concept of environmental sustainability and procurement. Some researchers have begun to face this issue, including Pasquire (1997) and Elliot and Palmer (1997). Elliot and Palmer considered which procurement system was best able to accommodate environmental assessment, concluding that the traditional system or construction management (which both give the client a high degree of control) or partnering would be most effective. To ensure that the work in these two areas begins to converge it is important that commission members not only engage with these concepts but also engage with colleagues from the CIB task groups working in this area.

Conclusions

As of yet it has been seen that there is no commonly acceptable definition of construction procurement. However, the wide-ranging subject matter, which has been brought to W92 by researchers and practitioners, has moved substantially away from any simple definition concerning, for example, client choice or contractual arrangements. Authors have found it necessary to define what they mean by construction procurement but invariably a strict application of one definition would constrain valuable work by other researchers.

Much of the academic and technical press is not helpful to practitioners who are seeking to share their own experiences or to benefit from the experience of others. From a cursory glance at some procurement press and literature it is not easy to understand how construction is 'brought about'. Terms such as 'procurement systems', 'partnering', 'new production philosophies', 'total quality management', 'client satisfaction', 'concession contracts', and 'BOOT projects' are used to describe policies, practices or concepts interchangeably. For example, the partnering literature frequently describes a method, commonly generic for application by any client in any circumstance, to the extent that the term has become almost unusable.

For efficient transfer of knowledge the concepts and principles behind partnering need to be exposed and distinguished from the methods or tools; appropriate methods or tools can be developed by practitioners to suit the circumstances. The role that CIB W92 can play is to encourage researchers to expose and explore their conceptual frameworks and to support practitioners in conceptualising their problems. W92 should continue to provide the forum for the exchange of methodologies as well as the publication of research results.

Through the identification of the strategic issues in the work of CIB W92 to date this chapter has shown how the Commission has already made significant progress towards meeting its objectives. Although there has long been some evidence that procurement systems are not purely technical and organisational subsystems but are a much more complex social interaction, the role for W92 is to ensure that research and practice does not neglect this latter aspect. The essence of this paradigm is that, although structure is an aspect of procurement, it is inevitable that the social interactions that exist in project teams are a key determinant of success.

Much of the work of W92 complements and is complementary to the work of other CIB Commissions or task groups. This is particularly true in the areas of organisation and management, culture in construction, environmental sustainability, building economics, conflict in construction and construction in developing countries. This breadth is an indicator of the cross-disciplinary influences within CIB W92. Wide-ranging subject areas are covered; the Commission should recognise this as a strength but should also seek to focus on the manifestation of the phenomena being studied within the context of procurement and contractual relationships in construction. For example, the study of the significant role that governments can play even in market economies, the comparison and benchmarking of the procurement regulations of the various trading blocks and international agencies and researching and debating roles for institutions charged with construction industry development are all legitimate areas of work for the Commission.

References

Abdel-Meguid, T. and Davidson, C. (1996) 'Managed claims procurement strategy (MCPS): a preventive approach' in R. Taylor (ed.) *North meets south: proceedings of CIB W92 Procurement Systems Symposium*, University of Natal, Durban, 11–20.

Akintoye, A. and Taylor, C. (1997) 'Risk prioritisation of private sector finance of public sector projects, in C.H. Davidson and T.A. Abdel Meguid (eds) *Procurement – a key to innovation: proceedings of CIB W92 Symposium*, Université de Montréal, Montréal, CIB publication 203, 1–10.

Alarcon, L. (ed.) (1997) *Proceedings of International Conferences on Lean Production*, Rotterdam: Balkema.

Allin, S. (1990) 'Registration of contractors', in B. Bosanac (ed.) *International Symposium on Procurement Systems*, Gradevinski Institut, Zagreb, CIB publication 132, 8.

Alsagoff, A. and McDermott, P. (1994) 'Relational contracting: a prognosis for the UK construction industry', in S.M. Rowlinson (ed.) *East meets west: proceedings of CIB W92 Procurement Systems Symposium*, University of Hong Kong, Hong Kong, CIB publication 175, 11–19.

Aziz, A. and Ofori, G. (1996) 'Developing world-beating contractors through procurement policies: the case of Malaysia', in R. Taylor (ed) *North meets south: proceeding of CIB W92 Procurement Systems Symposium*, University of Natal, Durban, 1–10.

Baxendale, T., Dulaimi, M. and Tully, G. (1996) 'Simultaneous engineering and its implications for procurement', in R. Taylor (ed.) *North meets south: proceedings of CIB W92 Procurement Systems Symposium*, University of Natal, Durban, 21–30.

Bowen, P.A., Hindle, R. D., and Pearl, G. (1997) 'The effectiveness of building procurement systems in the attainment of client objectives', in C.H. Davidson and T.A. Abdel Meguid (eds) *Procurement – a key to innovation: proceedings of CIB W92 Sympoisum*, Université de Montréal, Montréal, CIB publication 203, 39–49.

Brochner, J. (1989) 'Building procurement – key to improved performance', in D. Cheetham, D. Carter, T. Lewis and D.M. Jaggar (eds) *Contractual procedures for building: proceedings of the International Workshop*, 6–7 April, University of Liverpool, Liverpool, UK, 83.

Carter, D.J. (1990) 'The use of the concept of activity profiles in contractual systems', in B. Bosanac (ed.) *International Symposium on Procurement Systems*, Gradevinski Institut, Zagreb, CIB publication 132, 30–9.

Chau, K.W. and Walker, A. (1994) 'Institutional costs and the nature of subcontracting in the construction industry', in S.M. Rowlinson (ed.) *East meets west: proceedings of CIB W92 Procurement Systems Symposium*, University of Hong Kong, Hong Kong, CIB publication 175, 371–8.

Cheung, S.O. (1997) 'Planning for dispute resolution in construction contracts', in C. H. Davidon and T.A. Abdel Meguid (eds) *Procurement – a key to innovation: proceedings of CIB W92 Symposium*, Université de Montréal, Montréal, CIB publication 203, 71–80.

Conlin, J.T., Langford, D.A. and Kennedy, P. (1996) 'The relationship between construction procurement strategies and construction disputes', in R. Taylor (ed.) *North meets south: proceedings of CIB W92 Procurement Systems Symposium*, University of Natal, Durban, 66–82

Cottrell, G.P.(1989) 'Sub-contracting', in D. Cheetham, D. Carter, T. Lewis and D.M. Jaggar (eds) *Contractual procedures for building: proceedings of the International Workshop*, 6–7 April, University of Liverpool, Liverpool, UK, 117.

Craig, R. (1996) 'International public procurement systems', in R. Taylor (ed.) *North meets south: proceedings of CIB W92 Procurement Systems Symposium*, University of Natal, Durban, 83–92.

Davenport, D. (1994) 'Assessing the efficiency of international procurement systems in order to improve client satisfaction with construction investment – the French experience', in S.M. Rowlinson (ed.) *East meets west: proceedings of CIB W92 Procurement Systems Symposium*, University of Hong Kong, Hong Kong, CIB publication 175, 43–51.

Davidson, C.H. (ed.) (1998) *Procurement – the way forward: proceedings of CIB W92 Symposium*, Université de Montréal, Montréal, CIB publication 203.

Dimkic, Z. (1990) 'Some aspects of Yugoslav contractual arrangements and comparison with international practice in the construction industry', in B. Bosanac (ed.) *International Symposium on Procurement Systems*, Gradevinski Institut, Zagreb, CIB publication 132, 52–8.

Doree, A. (1991) 'How professional clients obtain design', in *Procurement Systems Symposium*, Las Palmas, Gran Canaria, Spain, CIB publication NI 145, chapter 11.

Dreger, G.T. (1993) 'Design–build procurement: strategies for success'. in T. M. Lewis (ed.) *Proceedings of CIB W65 Symposium*, University of the West Indies, Port of Spain, Trinidad, vol. 1. 749–58.

Dulaimi, M.F. and Dalziel, R.C. (1994) 'The effects of the procurement method on the level of management synergy in construction projects', in S.M. Rowlinson (ed.) *East meets west: proceedings of CIB W92 Procurement Systems Symposium*, University of Hong Kong, Hong Kong, CIB publication 175, 53–59.

Eastham, R.A. (1990) 'The decision to tender within current contractual arrangements', in B. Bosanac (ed.) *International Symposium on Procurement Systems*, Gradevinski Institut, Zagreb, CIB publication 132, 68–78.

Edwards, P. and Bowen, P. (1996) 'Building procurement in the "new" South Africa: the communication imperative', in R. Taylor (ed.) *North meets south: proceedings of CIB W92 Procurement Systems Symposium*, University of Natal, Durban, 120–9.

Egbu, C., Torrance, V. and Young, B. (1996) 'The procurement of project works in the ship refurbishment and construction industries', in R. Taylor (ed.) *North meets south: proceedings of CIB W92 Procurement Systems Symposium*, University of Natal, Durban, 130–40.

Elliot, C. and Palmer, A. (1997) 'Measuring sustainability using traditional procurement systems', in C.H. Davidson and T.A. Abdel Meguid (eds) *Procurement – a key to innovation: proceedings of CIB W92 Symposium*, Université de Montréal, Montréal, CIB publication 203, 175–85.

Fellows, R. (1989) 'Development of British building contracts', in D. Cheetham, D. Carter, T. Lewis and D.M. Jaggar (eds) *Contractual procedures for building: proceedings of the International Workshop*, 6–7 April, University of Liverpool, Liverpool, UK, 14–19.

Fenn, P. (1989) 'Settling disputes in construction projects', in D. Cheetham, D. Carter, T. Lewis and D.M. Jaggar (eds) *Contractual procedures for building: proceedings of the International Workshop*, 6–7 April, University of Liverpool, Liverpool, UK, 79.

Fukuyama, F. (1995) *Trust: the social virtues and the creation of prosperity*, London: Hamish Hamilton.

Gidado, K. (1996) 'Political and economic development in Nigeria, what procurement system is suitable?', in R. Taylor (ed.) *North meets south: proceedings of CIB W92 Procurement Systems Symposium*, University of Natal, Durban, 160–8.

Gow, H.A. and Fenn, F.P. (1989) 'The UK experience: Scotland', in D. Cheetham, D. Carter, T. Lewis, and D.M. Jaggar (eds) *Contractual procedures for building: proceedings of the International Workshop*, 6–7 April, University of Liverpool, Liverpool, UK, 61.

Green, S. (1994) 'Sociological paradigms and building procurement', in S.M. Rowlinson (ed.) *East meets west: proceedings of CIB W92 Procurement Systems Symposium*, University of Hong Kong, Hong Kong, CIB publication 175, 89–97.

——— (1997) 'Rhetoric and reality: a social constructivist research agenda for business re-engineering in construction', in C.H. Davidson and T.A. Abdel Meguid (eds) *Procurement – a key to innovation: proceedings of CIB W92 Symposium*, Université de Montréal, Montréal, CIB publication 203, 203–12.

Gunning, J.G. and McDermott, M.A.(1997) 'Development in design and build contract practice in Northern Ireland', in C.H. Davidson and T.A. Abdel Meguid (eds) *Procurement – a key to innovation: proceedings of CIB W92 Symposium*, Université de Montréal, Montréal, CIB publication 203, 213–22.

Hamilton, N. (1990) 'A review of UK project procurement methods', in B. Bosanac (ed.) *International Symposium on Procurement Systems*, Gradevinski Institut, Zagreb, CIB publication 132, 100–9.

Hashim, M. (1997) 'Clients' criteria on the choice of procurement systems – a Malaysian experience', in C.H. Davidson and T.A. Abdel Meguid (eds) *Procurement – a key to innovation: proceedings of CIB W92 Symposium*, Université de Montrèal, Montréal, CIB Publication 203, 273–85.

Heath, B. and Berry, M. (1996) 'An examination of the issues arising from the use of standard and non standard procurement methods and the implications of same for project success', in R. Taylor (ed.) *North meets south: proceedings of CIB W92 Procurement Systems Symposium*, University of Natal, Durban, 200–12.

Hibberd, P. (1991) 'Key factors in procurement', in *Procurement Systems Symposium*, Las Palmas, Gran Canaria, Spain, CIB publication NI 145, Chapter 8.

Hibberd, P., Basden, A., Brandon, P.S., Brown, A.J., Kirkham, J.A.J. and Tetlow, S. (1994) 'Intelligent authoring of contracts', in S.M. Rowlinson (ed.) *East meets west: proceedings of CIB W92 Procurement Systems Symposium*, University of Hong Kong, Hong Kong, CIB publication 175, 115–24.

Hill, F. (1991) 'Urban development corporations in England and Wales', in *Procurement Systems Symposium*, Las Palmas, Gran Canaria, Spain, CIB publication NI 145, Chapter 5.

Hodgson, G. (1994) 'Cross border trading through contractor design: the United Kingdom/European community highway procurement model', in S.M. Rowlinson (ed.) *East meets west: proceedings of CIB W92 Procurement Systems Symposium*, University of Hong Kong, Hong Kong, CIB publication 175, 125–31.

Hughes, W. (1990) 'Designing flexible procurement systems', in B. Bosanac (ed.) *International Symposium on Procurement Systems*, Gradevinski Institut, Zagreb, CIB publication 132, 110–19.

Hutton, W. (1994) *The state we're in*, London: Jonathan Cape.

Ireland, V. (1984) 'Virtually meaningless distinctions between nominally different procurement systems', in *CIB W65 Proceedings of the 4th International Symposium on Organisation and Management of Construction*, University of Waterloo, Waterloo, Ontario.

Jackson, P.M. and Price, C.M. (1994) *Privatisation and regulation: a review of the issues*, London: Longman.

Jennings, I. and Kenley, R. (1996) 'The social factor of project organisation', in R. Taylor (ed.) *North meets south: proceedings of CIB W92 Procurement Systems Symposium*, University of Natal, Durban, 239–50.

Klimov, V.A., Didkovski, V.M. and Rekitar, Y.A. (1990) 'Contracts and tenders in construction in the USSR', in B. Bosanac (ed.) *International Symposium on Procurement Systems*, Gradevinski Institut, Zagreb, CIB publication 132, 121–34.

Kumaraswamy, M.M. (1994a) 'Growth strategies for less developed construction industries', in *10th Annual Conference of the Association of Researchers in Construction Management*, Loughborough University, Loughborough 154–63.

—— (1994b) 'New paradigms for procurement protocols', in S.M. Rowlinson (ed.) *East meets west: proceedings of CIB W92 Procurement Systems Symposium*, University of Hong Kong, Hong Kong, CIB publication 175, 143–8.

Lahdenpera, P. (1994) 'Increasing use of the product conception and its incorporation into various procurement systems', in S.M. Rowlinson (ed.) *East meets west: proceedings of CIB W92 Procurement Systems Symposium*, University of Hong Kong, Hong Kong, CIB publication 175, 149–58.

—— (1996) 'Re-engineering the construction process – formulation of a model-based research approach', in R. Taylor (ed.) *North meets south: proceeding of CIB W92 Procurement Systems Symposium*, University of Natal, Durban, 275–86.

Lam, P. and Chan, A.P.C. (1994) 'Construction management as a procurement method: a new direction for Asian contractors', in S.M. Rowlinson (ed.) *East meets west: proceedings of CIB W92 Procurement Systems Symposium*, University of Hong Kong, Hong Kong, CIB publication 175, 159–68.

Latham, M. (1994) *Constructing the team, Joint review of procurement and contractual arrangements in the United Kingdom construction industry: final report*, London: The Stationery Office.

Lavers, A. (1991) 'The implementation of the EC construction products directive in the UK', in *Procurement Systems Symposium*, Las Palmas, Gran Canaria, Spain, CIB publication NI 145, Chapter 7.

Lenard, D. and Mohsini, R. (1998) 'Recommendations from the organisational workshop', in C.H. Davidson (ed.) *Procurement – the way forward: Proceedings of CIB W92 Montrèal Conference*, Université de Montréal, Montréal, CIB publication 203, 79–81.

Leung, H.F., Chau, K.W. and Ho, D.C.W. (1996) 'Features in the Hong Kong airport core project's general conditions of contract – a comparison with the Hong Kong Government general conditions of contract for civil engineering works', in R. Taylor (ed.) *North meets south: proceedings of CIB W92 Procurement Systems Symposium*, University of Natal, Durban, 287–300.

Liu, A.M.M. (1994) 'From act to outcome – a cognitive model of construction procurement, in S.M. Rowlinson (ed.) *East meets west: proceedings of CIB W92 Procurement Systems Symposium*, University of Hong Kong, Hong Kong, CIB publication 175, 169–78.

Liu, A.M.M. and Fellows, R. (1996) 'Towards an appreciation of cultural factors in the procurement of construction projects', in R. Taylor, (ed.) *North meets south: proceedings of CIB W92 Procurement Systems Symposium*, University of Natal, Durban, 301–10.

Lloyd-Schut, W.S.M. (1991) 'Recent developments in EEC construction law', in *Procurement Systems Symposium*, Las Palmas, Gran Canaria, Spain, CIB publication NI 145, Chapter 6.

McCarthy, C. and Wang, W. (1994) 'A flexible knowledge based system for the new engineering contract', in S.M. Rowlinson (ed.) *East meets west: proceedings of CIB W92 Procurement Systems Symposium*, University of Hong Kong, Hong Kong, CIB publication 175, 195–202.

McDermott, P. (1996) 'Role conflict within the United Kingdom contracting system', in R. Taylor (ed.) *North meets south: proceedings of CIB W92 Procurement Systems Symposium*, University of Natal, Durban, 353–67.

McDermott, P. and Jaggar, D. (1991) 'Towards establishing the criteria for a comparative analysis of procurement methods in different countries: social systems in construction', in *Procurement Systems Symposium*, Las Palmas, Gran Canaria, Spain, CIB publication NI 145, Chapter 9.

—— (1997) 'A strategic exploration of procurement', unpublished paper presented at CIB W65 Conference, Glasgow.

McDermott, P., Melaine, Y. and Sheath, D. (1994) 'Construction procurement systems: what choice for the Third World?', in S.M. Rowlinson (ed.) *East meets west: proceedings of CIB W92 Procurement Systems Symposium*, University of Hong Kong, Hong Kong, CIB publication 175, 203–11.

McDermott, P., Rowlinson, S.M. and Jaggar, D. (1997) 'Foreword', in C.H. Davidson and T.A. Abdel Meguid (eds) *Procurement – a key to innovation: proceedings of CIB W92 Symposium*, Université de Montréal, Montréal, CIB publication 203, xv–xxii.

Martins, R. and Taylor, R. (1996) 'Cultural identities and procurement', in R. Taylor (ed.) *North meets south: proceedings of CIB W92 Procurement Systems Symposium*, University of Natal, Durban, late paper.

Merna, A. and Smith, N.J. (1991) 'Concessions and risks in BOOT projects', in *Proceedings of Association of Researches in Construction Management Annual Conference*, University of Bath, Bath, 11–19.

Mohamed, S. (ed.) (1997) *Construction Process Re-engineering: International Conference*, Gold Coast, Australia, 14–15 July.

Mohsini, R. and Davidson, C.H. (1989) 'Building procurement – key to improved performance', in D. Cheetham, D. Carter, T. Lewis, and D.M. Jaggar (eds)

Contractual procedures for building: proceedings of the International Workshop, 6–7 April, University of Liverpool, Liverpool, UK, 83.

Morgan, G. (1985) *Images of Organisation*, Beverly Hills, CA: Sage.

Mukalula, P. (1996) 'The effects of the structural adjustment programme (SAP) on maintenance procurement contracts (Zambia's case)', in R. Taylor (ed.) *North meets south: proceedings of CIB W92 Procurement Systems Symposium*, University of Natal, Durban, 419–29.

Mustapha, F.H., Naoum, S.G. and Aygun, T. (1994) 'Public sector procurement methods used in the construction industry in Turkey', in S.M. Rowlinson (ed.) *East meets west: proceedings of CIB W92 Procurement Systems Symposium*, University of Hong Kong, Hong Kong, CIB publication 175, 229–34.

Naoum, S.G. (1990) 'Management contracting – review and analysis', in B. Bosaanc (ed.) *International Symposium on Procurement Systems*, Gradevinski Institut, Zagreb, CIB publication 132, 135.

Naoum, S.G. and Coles, D. (1991) 'Procurement method and project performance', in *Procurement Systems Symposium*, Las Palmas, Gran Canaria, Spain, CIB publication NI 145, Chapter 10.

Naoum, S.G. and Mustapha, F.H. (1994) 'Influences of the client, designer and procurement methods on project performance', in S.M. Rowlinson (ed.) *East meets west: proceedings of CIB W92 Procurement Systems Symposium*, University of Hong Kong, Hong Kong, CIB publication 175, 221–8.

Ndekugri, I. (1989) 'Sub-contracting in the UK constructing industry', in D. Cheetham, D. Carter, T. Lewis, and D.M. Jaggar (eds) *Contractual procedures for building: proceedings of the International Workshop*, 6–7 April, University of Liverpool, Liverpool, UK, 139.

NEDO (1983) *Faster building for industry*, National Economic Development Office; London: The Stationery Office.

Newcombe, R. (1994) 'Procurement paths – a power paradigm', in S.M. Rowlinson (ed.) *East meets west: proceedings of CIB W92 Procurement Systems Symposium*, University of Hong Kong, Hong Kong, CIB publication 175, 243–50.

Ng, W.F. (1994) 'Procurement methods for rural housing projects in the poverty-stricken areas of Guangxi in the People's Republic of China', in S.M. Rowlinson (ed.) *East meets west: proceedings of CIB W92 Procurement Systems Symposium*, University of Hong Kong, Hong Kong, CIB publication 175, 251–8.

Ofori, G. and Pin, T. (1996) 'Linking project procurement to construction industry development: the case of Singapore', in R. Taylor (ed.) *North meets south: proceedings of CIB W92 Procurement Systems Symposium*, University of Natal, Durban, 473–82.

Ogunlana, S. (1997) 'Build–operate–transfer procurement traps: examples from transportation projects in Thailand', in C.H. Davidson and T.A. Abdel Meguid (eds) *Procurement – a key to innovation: proceedings of CIB W92 Symposium*, Université de Montréal, Montréal, CIB publication 203, 585–94.

Ogunlana, S. and Malmgren, C. (1996) 'Experience with professional construction management in Bangkok, Thailand', in R. Taylor (ed.) *North meets south: proceedings of CIB W92 Procurement Systems Symposium*, University of Natal, Durban, 483–91.

Pasquire, C. (1994) 'Early incorporation of specialist design capability', in S.M. Rowlinson (ed.) *East meets west: proceedings of CIB W92 Procurement Systems Symposium*, University of Hong Kong, Hong Kong, CIB publication 175, 259–67.

—— (1997) 'The implications of environmental issues on construction procurement', in C.H. Davidson and T.A. Abdel Meguid (eds) *Procurement – a key to innovation: proceedings of CIB W92 Symposium*, Université de Montréal, Montréal, CIB publication 203, 603–15.

Pollock, R.W. and Rees, K. (1989) 'The potential for expert systems in the interpretation of building contracts', in D. Cheetham, D. Carter, T. Lewis and D.M. Jaggar (eds) *Contractual procedures for building: proceedings of the International Workshop*, 6–7 April, University of Liverpool, Liverpool, UK, 73–82.

Potts, K. and Toomey, D. (1994) 'East and West compared: a critical review of two alternative payment systems', in S.M. Rowlinson (ed.) *East meets west: proceedings of CIB W92 Procurement Systems Symposium*, University of Hong Kong, Hong Kong, CIB publication 175, 269–76.

Potts, K. and Weston (1996) 'Risk analysis estimation and management on major construction works', in R. Taylor (ed.) *North meets south: proceedings of CIB W92 Procurement Systems Symposium*, University of Natal, Durban, 522–31.

Ramsay-Dawber, P.J. (1996) 'Business performance measures of the UK construction companies: an aid to pre-selection', in R. Taylor (ed.) *North meets south: proceedings of CIB W92 Procurement Systems Symposium*, University of Natal, Durban, 532–9.

Robinson, N.M. (1990) 'Towards a multi-purpose modular form of construction contract', in B. Bosanac (ed.) *International Symposium on Procurement Systems* Gradevinski Institut, Zagreb, CIB publication 132, 144–53.

Rowlinson, S.M. and Root, D. (1997) 'The impact of culture on project management', *Final Report*, HK/UK Joint Research Scheme, Department of Real Estate and Construction, Hong Kong.

Saito, T. (1994) 'The comparative study of procurement system in the UK and Japan', in S.M. Rowlinson (ed.) *East meets west: proceedings of CIB W92 Procurement Systems Symposium*, University of Hong Kong, Hong Kong, CIB publication 175, 389–401.

Sharif, A. and Morledge, R. (1994) 'A functional approach to modelling procurement systems internationally and the identification of necessary support frameworks', in S.M. Rowlinson (ed.) *East meets west: proceedings of CIB W92 Procurement Systems Symposium*, University of Hong Kong, Hong Kong, CIB publication 175, 295–305.

Singh, S. (1990) 'A rational procedure for the selection of appropriate procurement systems', in B. Bosaanc (ed.) *International Symposium on Procurement Systems* Gradevinski Institut, Zagreb, CIB publication 132, 175–82.

Smith, A. and Wilkins, B. (1994) 'Procurement of major publicly funded health care projects', in S.M. Rowlinson (ed.) *East meets west: proceedings of CIB W92 Procurement Systems Symposium*, University of Hong Kong, Hong Kong, CIB publication 175, 307–14.

Suite, W. (1996) 'The provision of infrastructure in developing countries', in D.A. Langford and A. Retik (eds) *The organisation and management of construction: proceedings of CIB W65 Symposium*, Glasgow, 2.

Swanston, R. (1989) 'United Kingdom procurement procedures', in D. Cheetham, D. Carter, T. Lewis, and D.M. Jaggar (eds) *Contractual procedures for building: proceedings of the International Workshop*, 6–7 April, University of Liverpool, Liverpool, UK, 23–36.

Tam, C.M., Li, W.Y. and Chan, A. (1994) 'BOT applications in the power industry of Southeast Asia: a case study in China', in S.M. Rowlinson (ed.) *East meets west: proceedings of CIB W92 Procurement Systems Symposium*, University of Hong Kong, Hong Kong, CIB publication 175, 315–22.

Taylor, R. and Norval, G. (1994) 'Developing appropriate procurement systems for developing communities', in S.M. Rowlinson (ed.) *East meets west: proceedings of CIB W92 Procurement Systems Symposium*, University of Hong Kong, Hong Kong, CIB publication 175, 323–34.

Tiong, R.L.K. (1990) 'BOT projects: risks and securities', *Construction Management and Economics*, 8: 315–28.

Torkornoo, G.A.E. (1991) 'Procuring construction in Ghana', in *Procurement Systems Symposium*, Las Palmas, Gran Canaria, Spain, CIB publication NI 145, 19.

Vukovic, S. and Marinic, I. (1994) 'Organizational, technological and economic development of construction companies under the conditions of privatization in Vojvodina', in S.M. Rowlinson (ed.) *East meets west: proceedings of CIB W92 Procurement Systems Symposium*, University of Hong Kong, Hong Kong, CIB publication 175, 335–42.

Wahlstrom, O. (1989) 'Simplified tender documents giving an unambiguous representation of the finished building', in D. Cheetham, D. Carter, T. Lewis, and D.M. Jaggar (eds) *Contractual procedures for building: proceedings of the International Workshop*, 6–7 April, University of Liverpool, Liverpool, UK, 107.

Walker, D.H.T. (1996) 'Characteristics of a good client's representative', in R. Taylor (ed.) *North meets south: proceedings of CIB W92 Procurement Systems Symposium*, University of Natal, Durban, 614–24.

Walker, D.H.T. (1994) 'Procurement systems and construction time performance', in S.M. Rowlinson (ed.) *East meets west: proceedings of CIB W92 Procurement Systems Symposium*, University of Hong Kong, Hong Kong, CIB publication 175, 343–51.

Williams, P. (1991) 'The new engineering contract', in *Procurement Systems Symposium*, Las Palmas, Gran Canaria, Spain, CIB publication NI 145, Chapter 4.

Williamson, O. (1975) *Markets and hierarchies: analysis of anti-trust implications: a study in the economics of internal organisation*, New York: The Free Press.

2   A definition of procurement systems

Steve Rowlinson

Definitions

It is important at the outset to ensure that the field of discussion is clearly defined. One of the problems that has beset this particular area of construction research is a lack of clear definitions as to what is meant by procurement systems or terms such as contract strategy, or even the meaning of design and build. Hence, the definitions used in this book are briefly discussed below; for a more detailed discussion of the definition of procurement systems please turn to Chapter 12.

The first issue to distinguish clearly is what is meant by construction. In this book we are looking at the whole project life cycle, from initial inception of the project to its realisation and use; this accords with Walker's (1996) systems view of the process. First, for reasons of simplicity of presentation, the project process is seen as being divided into three distinct processes: design, construction and use. Within the concept of design there is the whole range of planning, funding, structural and architectural design and documentation; in short all of those activities which are necessary in order to be able to break ground on a new site. The construction process is seen as involving all of those activities, be they technical, managerial or strategic, which make up the realisation phase of the project where the physical facility actually appears. On completion of this phase the facility is actually used and this is an important part of the whole process: the use phase of a project has a major impact on the client's perception of whether the process has been successful or not.

The perspective adopted in this book is multifaceted in that the client view of the process is presented as well as the contractor's view, particularly with reference to such issues as contract strategy and concepts such as partnering. Thus the overall perspective adopted is one of simplification of the phases of the construction process and the polarisation of the procurement system into two actors: the client and the construction industry. This is not to say that we will treat the subject areas superficially but what it does allow is a simplification of the model of the project process. Through this simplification a focus on specific elements can then be maintained.

The type of construction referred to is generic in the sense that we are dealing with all types of projects in the building, civil engineering and process industries. However, it has to be noted here that many of the examples used come from the building industry and so the insights offered are particularly focused on this sector.

The client

The client is the sponsor of the construction process. The nature of the client is discussed below but the key issue to bear in mind is that the client provides the most important perspective on how the construction industry performs as far as procurement systems are concerned. An example of the power of the client came with the introduction of the British Property Federation System in 1984. This provided an example of sophisticated clients attempting to force change upon the construction industry. The focus of Sir Michael Latham's report (1994) was the client and client expectations of the construction industry. Hence, parts of this book will focus on the client, the client role and client expectations of the industry. This is a significant part within the procurement system as illustrated in Figure 2.1.

Procurement systems

The discussion of procurement systems in a broad sense takes place throughout this book but it is necessary here to outline what such a system

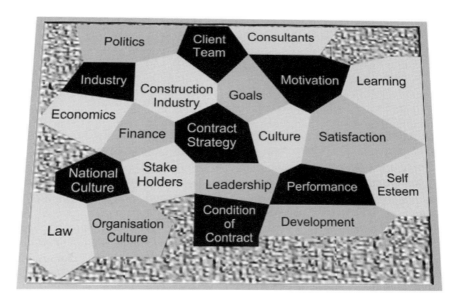

Figure 2.1 A systems view of procurement

entails. With reference to Figure 2.1, it is important to recognise that elements such as contract strategy and the client are functional parts of the procurement system. However, the effectiveness of the client organisation or the contract strategy is modified by other procurement system variables such as culture, sustainability, economic and political environment and more practical concepts such as partnering.

Contract strategy

Contract strategy is a key component of a procurement system. One of the problems besetting the study of procurement systems is the commonly held belief that a procurement system is defined by a simple contract strategy. An example of this is the current trend in the USA to sell design and build as the best solution for any construction project. This simple statement underscores two problems in studying this area. First, design and build does not in any way adequately describe a procurement system. The issue of definition of procurement systems has been dealt with many times and was well described by Ireland (1984) in a paper entitled 'Virtually meaningless distinctions between nominally different procurement forms'. Later in this chapter (pages 45–7) contract strategy is defined by a set of seven variables, but what must be borne in mind is that this view of contract strategy must then be put together with the whole procurement system with all the political, social and economic factors which impinge upon any project. The second problem with this simple statement is the view that a best solution exists. The view adopted throughout most of this work is that a contingency view pertains to procurement systems; that is, there are a series of different routes to a successful project all starting from the same point but following different courses. What has to be identified are all the key issues and participants in the procurement process before a suggested procurement system, and as part of this a contract strategy, can be devised.

The client in context

Historical background

At the simplest level the client can be regarded as the sponsor of the building process – the organisation that initiates the building process and appoints the building team. This client body has an identified need for a project but, generally, its business is not construction but some other process, be that manufacturing, sale or rental of property, commerce, retail, services or other productive processes. The change in the client body has been identified by Newcombe (1994) as a move from an individual to a corporation, from a unitary body or single person to a series of stakeholders. Walker and Kalinowski (1994) give a good example of the complexity of the client organisation in their paper describing the construction of the Convention and Exhibition Centre extension in Hong Kong.

Nature

At the simplest level of analysis both Masterman (1992) and Newcombe (1994) have described the client as being naive, having a need to purchase construction expertise from outside the organisation, or sophisticated, having appropriate skill and expertise within the organisation. In addition, Newcombe (1994) considers two client groups: public authorities, which are accountable to society for the projects undertaken and so have an obligation to provide some form of service, and private organisations with their main obligation being to maximise stakeholders' benefits to the exclusion of other objectives.

Legal standing

Legally, the client can be well-defined as, normally, a signature appears on the contract documents. However, in practice the client organisation may well be a whole series of competing and contrasting stakeholders. The client is certainly not a unitary body and may take the nature of a multi-organisation. This multiorganisation may be temporary or permanent. Understanding the nature of the client in this context and the legal and contractual and organisational obligations of the client is an important aspect of the procurement system. This issue is taken up in detail by Green and Lenard in Chapter 3.

Commercial standing

Commercially the client may exist in one of three broad areas:

- outside the construction industry, such as the Hongkong Bank or Glaxo plc;
- within the construction industry, such as government or local authority departments with their own professional staff;
- partially outside the construction industry, such as property developers with associated construction companies, for example New World Development or Slough Estates.

This position of the client will in some cases reflect the client's expertise in dealing with procurement systems, but it is obvious that a large organisation with a continuing development programme will be far better placed to deal with the construction industry than will a new company with no previous building experience.

Views of the client

A problem which affects many construction organisations is the impatience that they have with the client organisation; it is common for a separation of ownership and occupation of development projects to exist and this generates a situation where many stakeholders have an interest in the contruction project. Inevitably this leads to conflicts of interest in the specification of the facility. Cherns and Bryant deal with this issue when they state:

> Each participant (in the client team) can be seen as bringing to the table his [or her] own sense of what is at risk personally, as well as what is at stake professionally or departmentally, in the forthcoming project experience. . . . Many of the stakes are reputational. . . . in considering the role of the client, then, we cannot treat the client as unitary.
>
> (*idem* 1984: 140)

The question to be asked is, 'Who is it that actually meets the client in the procurement process?' Traditionally in the construction industry it has been the architect who has taken the client's brief and advised on the appointment of other consultants and contractors and subcontractors, and these other participants rarely meet with the client. Hence, the architect has become a surrogate client who participates in the transmission of information, with or without distortion, to the rest of the team. This phenomenon was highlighted by a number of researchers, including Wood in 1975. Owing to dissatisfaction with this situation a number of different procurement systems became popular. Each of these new systems has involved other construction team members moving closer to the client in an attempt to achieve a more successful project outcome. Examples of such systems are design and build, construction management and management contracting.

The view taken in this chapter is that the client has moved from being:

- an occasional builder to a regular builder;
- naive of the construction process to being highly sophisticated;
- a distinct person or body to a much more unfocused and temporary multiorganisation;
- outside the construction industry to within it.

In short, the 'client' has evolved over the past few decades and has questioned the ability of the construction professionals to counsel it effectively on its building need.

Client expectations

Objectives

Once one accepts that the client is not a unitary body it is then possible to start to build an individual profile for the project. If this profile is to be of use one must be able to extract from the client body the objectives that the client has in mind in going about the procurement process. These objectives will obviously vary from client to client but will include ideas such as:

- a minimum disruption to the client's ongoing business;
- an appropriate level of client involvement in the procurement process;
- a particular assessment, which is right for the individual client, of the importance of cost, time, quality, function and performance of the project.

Before attempting to advise a client on an appropriate contract strategy it is essential that the client's objectives be clearly defined and stated and tested within the client body. All clients require value for money but, without clear objectives, assessment of what is value for money is impossible. It is almost certain that most clients will look to see increased profitability and productivity from the outcome of the procurement process. It is simply not good enough for the construction industry to ignore these key objectives of the client. The recent incursion of management consultants into the construction project management field should alert construction professionals of the need to get close to their clients if they wish to maintain their competitiveness within the construction procurement process.

Expectations

In discussing the expectations that a client has of the construction industry Newcombe (1994) indicates that one of the most important things that a client expects from the construction industry is unbiased advice and accurate information. In addition to this the client requires protection from defaulting contractors and consultants and this is tied up with the need for a well-developed and clearly understood legal framework. The client also needs advice in identifying those elements of the procurement process which will bring about performance which matches the client's objectives. It is this specific area which this book addresses and, in the following chapters, the variables which need to be considered in this process will be enumerated and discussed.

Attitude

Some of the questions which the client must address and have answers to from the construction industry are concerning its levels of involvement in the procurement process and division of responsibility and authority between the client organisation and the construction team. Hand in hand with this the client must make a decision on the issue of risk-sharing in the procurement process. This is a complex issue and is at the heart of the consideration of appropriate contract strategy. The client must also be aware of the increased cost that stems from building flexibility (the opportunity to change the project parameters) into the procurement system. Flexibility has a price.

Role

Essentially, the client's role in the procurement system is a strategic role: the client body sets the objectives and the construction industry turns those objectives into reality. As such, the strategic decisions made at the very outset of a project are the most crucial and so procurement systems and contract strategy must be critically reviewed and evaluated by the client. It has been in the area of strategy that the construction industry has been particularly weak in the past and this has led to the development of alternative procurement systems and the encroachment of other professions into the construction industry. Walker elaborates on this strategic phase of the project process in his book *Project management in construction* (1996), and Newcombe (1994) discusses the strategic nature of the procurement system and Chandler's (1966) axiom of 'structure following strategy' as being the key to organisational design in the procurement process.

Industry expectations

In his discussion of the expectations of the construction industry of its clients, Newcombe (1994) points out that the most obvious thing that the industry expects from the client is the decision to build. This is not simply a one-off decision though; the industry is looking particularly for continuity of work from a repeating client. In the past this has resulted in the formulation of new procurement systems such as the relationship between Bovis and the retailer Marks and Spencer plc (MPBW 1970). These relationships have enabled the contractor to get closer to the client organisation whilst participating in an ongoing building programme. Such initiatives are not altogether common, however, and new alternatives, such as the concept of partnering, have come to the fore in recent years as an attempt to deal with such matters.

Central to the construction industry's perception of the client is the nature of the brief-taking process and the desire of the industry to deal with a single client representative. The industry it seems would like to have a unitary client and a clear, well-defined brief. However, the reality of the situation is that the client is normally a complex multiorganisation with a whole range of stakeholders bringing their vested interests to the table, and the brief is actually an evolving process rather than a cast-in-stone document. Hence, there is a mismatch between what the industry would like to see and what the client actually provides. On the other hand, this is mirrored by many clients who would wish to have a single point of responsibility to deal with in the construction process. The popularity of design and build systems is predicated on this concept. In fact, Latham (1994) puts forward a 'Code of practice' for this very issue.

Ultimately, of course, the construction industry will judge the performance of a client by more tangible criteria such as prompt payment, fair and open tendering systems, the experience and knowledge to understand the implication of design changes and the ability to provide prompt, timely decisions.

Contract strategy: procurement systems

Procurement is about the acquisition of project resources for the realisation of a constructed facility. This is illustrated conceptually in Figure 2.2, which was produced by the International Labor Office (1984). The figure clearly illustrates the construction project as the focal point at which a whole series of resources coalesce. Central to this model is the client's own resources that are supplemented by the construction industry participants, that is, the consultants and the contractors along with the suppliers and subcontractors. The model clearly illustrates the need for the acquisition of resources in order to realise the project. This acquisition of resources is part of the procurement system – note, they are only a part of the system. This part of the system can be referred to as the contract strategy, that is, the

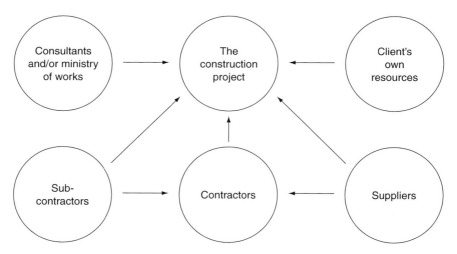

Figure 2.2 Procurement
Source: Austen and Neale 1984

process of combining these necessary resources together. The contract strategy is not the procurement system but only a part of it: the rationale behind this definition is that the procurement system involves other features such as culture, management, economics, environment and political issues.

Conventionally, contract strategies have been described as, for example, the traditional approach, construction management or build–operate–transfer. However, writers such as Ireland (1984) and Walker (1994) have indicated that there is little difference between many of these supposedly alternative strategies. This situation has occurred as a result of reluctance by many writers to delineate clearly the various variables which make up a contract strategy. The list given in Figure 2.3 indicates an increasing integration of design and construction expertise within the one organisation as one works down the list. However, to assume that these labels will uniquely define a contract strategy is a false supposition. What is needed is a set of key variables which can uniquely define a contract strategy rather than the arbitrary list of definitions given in Figure 2.3.

Why has this range of contract strategies developed within the construction industry? One of the reasons is illustrated in Figure 2.4: the vicious

Figure 2.3 Contract strategy

Figure 2.4 Vicious circles in construction procurement
Source: Curtis *et al.* 1991
Reproduced by kind permission of the Construction Industry Research and Information Association

circles in construction procurement. The logic behind the model is that each participant wishes to obtain as much as possible from the process in terms of its own financial rewards. Given the fact that the process takes place in a competitive market there is a cycle of fluctuating pressure on prices at all times. A downward pressure thus forces the contracting participant to look to alternative means to recoup its profit. This leads to claims-conscious behaviour and can also stimulate reductions in quality and functional performance. As a consequence the client and its advisers are forced into the position of exerting greater surveillance over the contractor in order to minimise the effects of this behaviour. As can be seen from the figure this results in a vicious cycle of negative behaviours. One way of avoiding this model, which is based around the traditional strategy, is to adopt alternative contract strategies.

Alternatively, other forces may be at work in bringing new contract strategies to the market place. The competitive nature of the market forces organisations to innovate if they wish to grow and secure market share. Innovation in procurement systems was the theme of the highly successful symposium held by the CIB's working party on procurement systems (CIBW92) in Montreal (Davidson and Tarek 1997). Hence, as the construction industry has typically been a conservative and somewhat traditional industry there has been, in the past, ample scope for the introduction of new strategies. However, until the late 1960s such attempts at innovation were quite rare, certainly in the UK construction industry. With the advent of more experienced and sophisticated clients there has been an opportunity for industry-leading contractors to explore new routes. Also, as buildings have become technically more complex and clients managerially more sophisticated there has been an increasing recognition that the conventional (traditional) approach to procurement is inadequate. Additionally, in recent years concepts such as business process re-engineering have taken root in client organisations, and the construction industry has experienced these concepts both at second hand, when working with a major client, and first hand, in the rush to reorganise in the face of declining markets. Consequently, factors have combined to force the construction industry into the position where it has to change to survive.

Organisation theory and the project process

Interdependence

A theme which comes out of the procurement systems literature is interdependence (Higgin *et al.* 1966; Wood 1975; Walker 1996). This interdependence is both in terms of the construction process and in terms of the organisational structure. It is typical of project-based organisations and is a key consideration in the analysis and design of the organisational structures and the relationships in the construction project team. In analysing and designing procurement systems this interdependence of both the organisations in the project team and of the tasks in the construction process present special problems which most client bodies have not faced before. As a consequence, the design of the project organisation presents unique problems for many construction client bodies and this, in part, can be seen as a cause of the discontent expressed with the performance of the construction industry by many clients. In this context, the choice of an appropriate procurement system takes on great significance.

Differentiation and specialisation

The construction industry is also characterised by high degress of specialisation, as evidenced by the many independent professions represented on a typical construction project and this specialisation inevitably leads to high degress of differentiation. Walker (1996) discusses this in the context of organisational design and identifies four levels of differentiation operating in the industry, namely, time, territory, task and sentient differentiation. This neatly sums up, in theoretical terms, the nature of the construction project team which is characterised by small groups of professionals working in different locations on specialist tasks which are all interdependent. The differentiation of the contractor from the rest of the team is most apparent and is highlighted by the different process of contractor appointment and payment compared with that for the consultants. Conventionally, the design and construction process, when broken down into its constituent parts, is seen as sequential and leading to the smooth flow of individual contributions through a project life cycle. This view is, of course, unrealistic and what happens in reality is that contributions are refined and revised and the new information is passed backwards for further analysis and redesign, described as a synchronous process. This recycling of work is a source of frustration and friction to participants who are working to the conventional model.

Integration and coordination

The necessary corollary to this specialisation of the contributors to the construction project is the process of integration and coordination. Integration may be achieved by adapting the organisational structure to fit the degree of differentiation present in the team. Examples of this in the construction industry are the appointment of a project manager who acts as an integrating device, linking team members together, or the use of design and build methods. Each of these is seen as a method of drawing the separate specialist functions of the construction process together and bringing down the barriers that exist between the separate organisations in the project team.

Walker (1996) describes the use of linear responsibility analysis as a means of assisting in organisational design; it is a technique which allows the separate technical and managerial roles to be clearly delineated and subsequently integrated. It is based on the systems theory precept that the managing system and the operating system should be separate systems. This is one area in which the traditional (conventional) system has been criticised in that the architect plays the roles of both designer and project manager, without any differentiation of roles.

Universal compared with contingency approaches

The search for the best procurement system has been the topic of many articles and papers over the years. However, if one adopts a contingency theory view of procurement systems then it is possible to identify a range of contingency factors which will impinge upon the effectiveness of the strategy chosen. Many researchers (for instance Walker 1994; Ireland 1983; NEDO 1983; Sidwell 1984) have identified contingency factors on construction projects and assessed their impact on project performance. Others have attempted to develop from such research a methodology for choosing a best contract strategy (NEDO 1985; Franks 1996; Skitmore and Marsden 1988). Liu (1995) has identified the fact that perceptions of project performance vary amongst members of the project team and within the client body. Hence, the problem of choosing a best procurement system becomes almost impossible as each participant will bring to the debate its own views and values, rendering the choice of a best system impossible. Further views on this subject can be found in Chapter 10, where Kumaraswamy and Walker discuss performance comparisons amongst procurement systems, and in Chapter 3, where Green and Lenard deal with the nature of the client. A major criticism of the attempts to produce a methodology for selecting a procurement system is that most methodologies deal with the selection of a (limited) contract strategy rather than dealing with the bigger issue of a procurement system. This is especially evident where an attempt is made to transfer a selection methodology from one country or region to another. Liu and Fellows discuss the impact of culture on procurement system effectiveness in Chapter 6, and Taylor *et al.* take up the theme of developmentally appropriate systems in Chapter 7.

Generic strategies

Much of the literature in this area uses terminology such as 'the traditional approach', 'design and build', 'build–operate–transfer', 'management contracting', etc. In order to define clearly the parameters being considered in this book a generic taxonomy of organisational forms is given below. The function of this taxonomy is to provide a clear and simple description of construction project organisational forms which, when taken with other contract strategy variables, uniquely define a strategy which is further clarified when put in the context of the overall procurement system.

Traditional approach

The traditional (or conventional) approach is shown diagramatically in Figure 2.5 and its key characteristic is the separation of the design and construction processes and the lack of integration across this boundary, along with the employment of a whole series of separate consultants to design the project and an independent contractor to take charge of the construction process. Typically, the project team will be led by an architect charged with the responsibility for both designing and managing the project. Other consultants will join the design and administration team (such as structural engineers and quantity surveyors) through the life of the project and the contractor will be selected from a competitive tendering process on a fixed price bid. The contractor's input to the design process will be minimal, often nil, and in many countries most of the production work on-site will be subcontracted to other organisations. The design and construction processes and their subtasks are seen as sequential and independent.

Design and build

The design and build approach is characterised by the single point responsibility offered to the client by the contractor and the opportunity for overlapping the design and construction phases which stems from this unitary approach (Figure 2.6). As with the traditional approach there are many

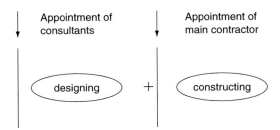

Figure 2.5 The traditional approach
Source: Austen and Neale 1984

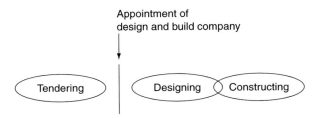

Figure 2.6 The design and build approach
Source: Austen and Neale 1984

variants on the basic theme of design and build. Of particular interest are the variants which include project financing and which go under the headings build–operate–transfer, build–own–operate–transfer, build–own–manage and the like. The organisation of a design and build project is more complex than that of the traditional project at the tender stage as the situation will often occur where different priced bids with different design solutions are competing for the same project. The adjudication of such bids is a complex process and requires, to be fair to all bidders, some assessment scheme to be put in place before bids are submitted.

Design and build projects go under many names, for example design–build, design and build, design–manage–construct, design and manage, build–operate–transfer (BOT), build–own–operate–transfer (BOOT), build–own–operate (BOO), turnkey, etc. The underlying principle behind all of these systems is that the client body contracts with one organisation for the whole of the design and construction process; the concept of single point responsibility exists. This, and the overlapping of the design and construction phases, are the underlying principles behind design and build systems. Hence, this can be considered as one of the three generic types of organisational forms in contract strategy and each form of design and build can be uniquely defined by addressing the status of the other contract strategy variables, such as leadership, selection process, payment systems, etc., as discussed below.

The use of design and build methods opens up competition to more than competition on price alone. The form of the finished project is also a part of the competition and so the process is much more open but also much more costly for the bidders. In such a situation it is necessary that the bidders be given some guarantee that the competition they are entering will be both fair and limited in terms of competitors. Prequalification of bidders is a virtual necessity for design and build projects.

Design and build organisations can be categorised into three main forms; this categorisation is based on the differentiation which each of these forms exhibits in terms of time, space and profession. The pure design and build organisation strives for a complete and self-contained construction system. All necesssary design and construction expertise resides within one organisation and this is sufficient to complete any task that arises. This organisation must specialise in a particular market sector because of the complexity of today's building projects. All aspects of design and construction can be highly integrated and much of the experience gained in design and construction is fed back into the organisation. Thus the potential for organisational learning in the design and build organisation is far greater than with other procurement systems (see Chapter 5 on organisational learning). There is a tendency, because of the need for market specialisation, for pure design and build organisations to stay within the medium-size range of contractors. Additionally, pure design and build organisations are unlikely to develop in countries with small construction markets; currently the USA has a large number of recently established design and build organisations which can take advantage of the size and scope of the US market.

The second form of design and build organisation is the integrated design builder. Such an organisation takes a less holistic approach to the design and construction team and buys in design or construction expertise whenever necessary. This may take the form of architectural or other consultancy services but a core of designers, engineers and project managers exists who are experienced in their own specialism and the workings of the organisation. These permanent staff provide the linking pin between the internal and external organisations and so exert an integrative influence on the team. The design and construction teams may be separate organisations within a business group and this group may participate in the whole range of contract strategies. This more general approach to design and build tends to be a development from a general contracting background and these organisations tend to be larger, more mature companies seeking particular market niches.

Finally, the simplest way for a construction organisation to enter the design and build market is to operate in a fragmented design and build mode. To perform in such a way the design group can be quite small, perhaps consisting solely of project managers whose task it is to liaise with the client and to appoint consultants to develop designs. Major companies have the ability to expand such units quite rapidly if required but, as with the traditional contract strategy, a major effort is required to integrate the work of the various consultants. This type of organisation is characterised by a lack of sense of identity and lack of feedback loops between the design and construction processes. The integration and coordination problems inherent in the traditional approach are likely to manifest themselves along with role ambiguity amongst the professions as they attempt to come to terms with working for a construction organisation acting as design team leader. Such fragmented design and build organisations have the capacity to take on large projects and such an approach is regularly used with BOT projects. The design and build forms are shown in Figure 2.7 (a more detailed discussion can be found in Rowlinson 1987).

Divided contract approach

The divided contract approach is illustrated diagramatically in Figure 2.8. The key principle in this form is the separation of the managing and operating systems. It can be clearly seen that the project organisation is over-arched by a managing system. This managing system is generally provided by a management contractor or a construction manager or a project manager. The tasks of design and construction are undertaken by separate organisations which specialise in the technical aspects of the process and their inputs are integrated and coordinated by the management organisation. The high degree of specialisation allows for the fast-tracking of the

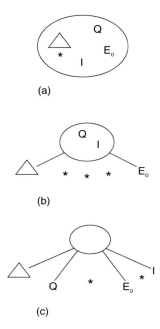

Figure 2.7 The design and build organisational types: (a) pure design and build; (b) integrated design and build; (c) fragmented design and build

Note: ○ = builder; Q = quantity surveyor; E_0 = services engineer; I = structural engineer; △ = architect; * = subcontractor

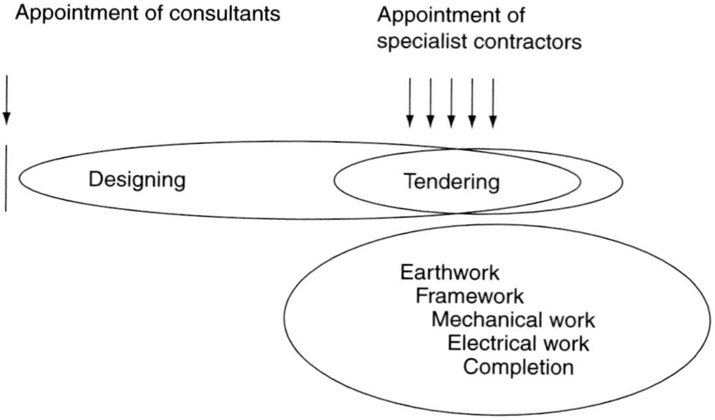

Appointment of consultants

Appointment of specialist contractors

Designing

Tendering

Earthwork
Framework
Mechanical work
Electrical work
Completion

Figure 2.8 The divided contract approach
Source: Austen and Neale 1984

project which is the fundamental characteristic of this type of organisational form.

The nature of this type of project organisation, and the underlying goal of providing a specialist management role, necessitates that the managing organisation is appointed at the outset of the project and the role of the contractor is one of consultant rather than constructor. This change in role leads to a reshaping of the roles that all of the other professionals play and so there is scope for role ambiguity and conflict to arise. Hence it is essential that roles and reporting relationships are clearly defined at the outset and that those consultants providing technical services are fully aware of the limits of their authority and responsibility. This management role is increasingly being undertaken by management consultants and this is a threat to the traditional construction industry participants that may find themselves distanced from their client base if this trend continues.

A comparison of US and UK practice [as defined by the Construction Industry Research and Information Association (CIRIA); Curtis *et al.* 1991] is provided in Figure 2.9 which illustrates the subtle difference between construction management and management contracting. The essential difference between these two organisational forms is the relationship of the client to the trades contractors. In the construction management system the client contracts directly with the trades contractors, and the construction manager (CM) has no contractual relationship with these contractors. The CM acts as the client's representative and has authority to make decisions affecting the designers and the contractors. In the management contracting system, on the other hand, the trades contractors have a direct contract with the management contractor (MC) and so the client is 'buffered' from the producers of the constructed project. The role of the MC is less powerful

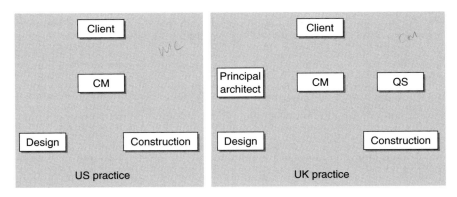

Figure 2.9 A comparison of US and UK practice
Source: Curtis *et al. 1991*
Note: QS = quantity surveyor; CM = construction manager
Reproduced by kind permission of the Construction Industry Research and Information Association

in that the MC does not necessarily have direct authority over the design consultants.

In essence, these two representations are idealisations of a complex, temporary multiorganisation and each project will be organised in a slightly different manner from the next. The nature of the payment system, legal documents and selection system will define more clearly the roles of each participant and so the exact nature of the contract strategy. In addition, the personalities involved, particularly of the different organisation leaders and their power bases, will have a significant effect on how the project organisation acts in practice (see for example, Walker and Newcombe, 1999).

The use of trades contractors in this approach allows the fast-tracking of the project as works packages are being let on a continuous, rolling programme as the design is developed. Although this leads to fast construction a price has to be paid for this in terms of abortive work and re-work as, for example, foundations may well be over designed to enable a rapid start on site (for an example, see Moss 1994). This strategy does encourage the use of value management techniques to improve design solutions, and on large projects, such as Broadgate in London (see Chapter 4), the opportunity to build a large integrated team at the site exists. The use of work packages also enables competitive tendering to take place for all elements of the construction process and so, although an extra management 'fee' has to be paid, a level of competition is maintained. Given the fast-track characteristics of this strategy and the extra costs associated with an overarching layer of management this approach has, conventionally, not been recommended for non-complex projects.

Figure 2.10 illustrates the range of management systems and their degree of professionalisation. The spectrum indicates the MC or CM acting as a consultant builder at one end of the spectrum and as a contractor at the other, that is, mercantilism compared with professionalism. In fact, the spectrum really illustrates two separate strategies, the right-hand side being, effectively, design and build, as one organisation takes sole respon-

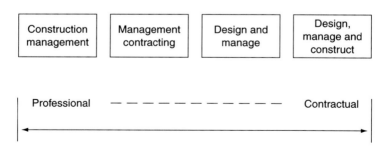

Figure 2.10 The range of management systems and their professional contract
Source: Curtis *et al.* 1991
Reproduced by kind permission of the Construction Industry Research and Information Association

sibility for the whole construction process, and, at the left-hand side the principle of separation of the design and management systems, typical of the divided contract approach, exists.

Contract strategy variables

It should now be clear from the foregoing discussion that the organisational form adopted cannot uniquely define the contract strategy being used. This takes us back to Ireland's concept of 'virtually meaningless distinctions between nominally different procurement forms'. In order to define contract strategy more clearly a minimum of seven separate variables must be considered. These are identified in the following subsections.

Organisational form

This variable has been discussed in detail above and defines the responsibilities of each of the disciplines in the project life cycle; whether they are directly responsible for making decisions or only expected to give advice and at what stage in the project life cycle they should be involved. These generic models allow only a partial evaluation of the construction process but do give a starting point for further definition.

Payment methods

Payment methods, particularly for contractors, are either cost-based or price-based. The latter method places more risk in the hands of the contractor whereas the former method places more risk at the feet of the client. Veld and Peeters (1989), in their paper 'Keeping large projects under control: the importance of contract type selection', discuss seven different payment methods and indicate the advantages and disadvantages of each. These, when combined with organisational form, give a reasonable feel for the risks involved in a particular contract strategy. Few standard contracts contain direct incentive provisions for the contractor to complete the contract early or for a cheaper price. If completion of sections of the work and overall completion is critical a payment system linked to milestones within the agreed construction programme can be used. Cost reimbursement and target contracts can also provide incentives to the contractor for reducing costs.

Overlap of project phases

This variable defines the degree of acceleration or fast-tracking that is desired within the construction process. In most instances it would be true to say that the traditional process is difficult to fast-track and little overlap occurs with this organisational form, but varying degrees of overlap occur with both the design and build and the divided contract approaches. The decision to overlap project phases also affects the selection process (both for contractors and for consultants). Conventional wisdom has it that the divided contract approaches can achieve the maximum overlap, with design and build also being able to achieve relatively high degrees of overlap and, consequently, fast project times.

Selection process

Contractors may be selected by open competition, by select competition among an unlimited number of prequalified contractors or by negotiation in one or more stages. This aspect of contract strategy is discussed in detail in Chapter 10. Similar methods may be adopted for the selection of consultants. This variable is linked both to payment methods and to the overlap of project phases in that open competition is normally used with price-based bids, and projects which need to be fast-tracked might use select competition. However, any of these processes can be used with any of the organisational forms but conventionally a set pattern has been implemented with the traditional system adopting open or select competition and design and build moving towards select competition or negotiation in the selection process. There is no reason to believe that conventional wisdom is correct.

Source of project finance

The source of project finance can have a significant impact on the contract strategy and procurement system chosen. If the client body provides finance then it essentially has a free hand in the choice of strategy, but if third parties or the contractor organisation provide part or all of the finance then strings will be attached. With aid agencies operating in non-industrialised countries there is a tendency to specify the use of funding-country contractors and products and this can have a negative effect upon the local construction industry (see Kumaraswamy and Dissanayaka, 1997). With contractor finance there is a tendency for this to be provided only when a franchise or other similar BOT-type agreement is incorporated into the project.

Contract documents

The contract documents used in construction projects are, in the main, drafted by the clients or industry bodies with representatives from all parties. The drafter will obviously weight the conditions towards its own interests but this is not necessarily a problem as long as the contract is well understood and disputes adequately documented. More important is the appropriateness of the documents to the type of contract strategy being used. Issues such as the degree of completion of drawings at commencement of construction and the use or not of bills of quantities or schedules of rates are important considerations. The contract documents should match the strategy and procurement system adopted. Most standard contracts are suitable only for a particular contract strategy but documents such as the Engineering and Construction Contract [previously known as the New Engineering Contract, (NEC)] produced by the UK Institution of Civil Engineeers is flexible enough to be used with all types of contract strategy.

Leadership

An important strategic decision is the choice of the project team leader. Any of the project participants, including the client, can take up this role. The choice of leader should be based on a number of factors, including personality, expertise and experience and an analysis of the roles and responsibilities to be allocated to the participants. In the past, in the traditional system, the role of leader went to the architect by default. If a contingency view of contract strategy is adopted then this choice by default must be questioned and the leader fitted to the strategy adopted.

Authority and responsibility

The distribution of authority and responsibility are important issues in any organisation and the design of the distribution of these is of paramount importance in project organisations, which have the characteristics of temporary multiorganisations. Walker (1996) discusses the nature of authority and responsibility in detail and suggests a methodology for devising an appropriate structure of authority levels within project organisations. Walker sees this as the key to success in poject organisation.

Performance

The performance of different organisational forms, and indeed procurement systems, has been the subject of much research over the past 20 years. No definitive outcome has stemmed from this research but a series of commonly held beliefs is presented in Table 2.1. The reader should be aware that comparison of the performance of the different organisational forms is fraught with difficulty and the opinion expressed cannot be relied upon in all circumstances.

Flexibility

Choice of an appropriate contract strategy is a difficult and complex process but one which many clients, particularly public sector clients, do

Table 2.1 Hypothesised performance of organisational forms

	Traditional	*Design and build*	*Divided contact*
Speed	L	H	H
Cost	M	L	H
Potential for incorporating variations	H	L	H
Cost certainty	M	M	L
Client involvement	L	M	H

Note: L = low; M = medium; H = high

not treat with sufficient care. It is also an old problem: for example, in the UK the following reports have been published which deal with the subject:

1944	Simon report,
1962	Emerson report,
1964	Banwell report,
1975	Wood report,
1983	Faster building for industry,
1983	Thinking about building,
1987	Faster building for commerce,
1994	Latham report,
1998	Egan report.

An effective contract strategy is one which is appropriate to the client, its organisation and the project itself. However, flexibility brings with it the need for control and the devising of appropriate procedures for choosing or drafting contract documents, selection procedures and payment procedures. Also, a price may have to be paid for choosing different strategies. For example, it can be argued that:

- a design and build approach involving overlapping of the design and construction phases leads to

  - a fast project,
  - an economical project,

  but that the client

  - has to make early design decisions,
  - must limit variations to a minimum,
  - finds evaluation of bids difficult;

- the use of variants on the management approach leads to

- a fast project,
- a high-quality product,

but that the client

- may incur extra management costs,
- must evaluate alternative services offered by bidders,
- takes a risk in proceeding with construction with incomplete designs.

The fallacy of cost certainty

Adopting the traditional approach based on full drawings and bills of quantities should give the client a firm, fixed price for construction, but in practice very few projects are actually completed within the tendered price. Indeed, the crux of the matter is that full drawings and a complete bill are often not available when the project goes to tender. This being the case, why do many organisations continue to use this method when it can be argued that it leads to:

- a lack of flexibility,
- a price to pay in terms of claims-conscious behaviour,
- the fallacy of cost certainty,
- a release of control by the client organisation?

Comparisons

A number of authors have attempted to characterise procurement systems (Masterman 1992; Turner 1990; Franks 1991, 1996) and from this characterisation many have attempted to derive a set of advantages and disadvantages for each system. The problem with such an approach is twofold; first, the definition of procurement system adopted is different in each case and is really more a description of contract strategy or organisational form rather than procurement system and, second, the pros and cons are viewed as absolutes rather than from a contingency theory viewpoint (this second criticism stemming in part from the view adopted in the first criticism).

For completeness, Table 2.2 presents the perceived advantages of different organisational forms as generally described in the literature. Such perceptions are important to note, as many of the procurement selection methodologies that have been produced have been based on a similar set of perceptions. The aim of this book is to dig beneath this set of perceptions to the more complex structure which constitutes procurement systems and so present a more detailed and complex model of the procurement process which identifies those factors critical to project procurement success.

As can be seen from Table 2.2 some of the advantages could be seen as

Table 2.2 The perceived advantages of three different organisational forms; the traditional approach, design and build, and management systems

Advantage	Disadvantage
Traditional approach:	
Competitive tendering is used	Decision processes are slow and
Bills of quantities make for ease of	convoluted
valuation of variations	Total project time is the longest of all
High quality and functional standards	options
are possible	It has low levels of buildability
There is cost certainly at the start of	Many organisational interfaces must
construction	be managed
Independent advice is given on most	
aspects of the process;	
There are clear lines of accountability	
A combination of the best design and	
construction skills is possible	
There is flexibility for design changes	
Design and build:	
There is single point responsibility	There is a lack of independent advice
Fixed price bids are used	Valuation of variations is not based on
Design and construction are integrated	fixed rates
It has high levels of buildability	It requires a detailed brief and client
Total project times are short	requirements at the outset
Client involvement in the process can be	Changes can be expensive
minimised	Client control of quality and
Package deal systems offer off-the-shelf	functionality is minimised
solutions	Design and build firms often lack
There is competition on price and	broad experience or expertise
product	There are lower levels of competition
	at tender than in the other approaches
	The tender process can be expensive to
	bidders
	Comparison of bids can be
	complicated
	It may not produce well-thought-out
	bids
Management system:	
It provides early start to construction	There is no firm price at the start of
work	construction
Quick total project times are possible	It can lead to complicated contractual
Value management and buildability are	relations and assignment of liability
emphasised	The extra layer of management leads
There is flexibility for change and	to extra costs
last-minute decisions	The client must be deeply involved at
There are high levels of competition for	most stages of the process
each work package, hence there is	It is conducive to 'buck-passing'
competitive tendering	More risk is taken by the client
The extra layer of management leads to	Conflict between professional
strong organisation and in-depth	participants may occur as a result of
planning of the whole process	changing roles
There is increased specialisation of	
trades contractor inputs	

disadvantages, and vice versa, depending on one's perspective. This observation serves to underline that a contingency view of the construction procurement process offers a sound basis for appraising the suitability of procurement systems and that procurement systems must be viewed in their entirety, not as separate elements (such as organisational form or contract strategy).

Conclusions

This chapter has attempted to present an overview of work in the area of procurement systems. The chapter has dealt with a number of issues relating to the definition of procurement systems and has highlighted the different ways in which procurement systems have been categorised in the past. The tendency to concentrate on organisational form or contract strategy has led to the omission of a number of important variables in the process of procurement system selection and analysis of performance. Thus, the groundwork of common definitions has been laid for use in the rest for this book. The following chapters will deal with particular aspects of the procurement system and the whole will be summarised in the concluding chapters. This chapter has made it clear that procurement systems are indeed complex systems and cannot be dealt with in a simple, straightforward manner. A sophisticated and holistic methodology needs to be identified in order to design and use procurement sytstems effectively.

References

Austen, A.D. and Neale, R.H. (eds) (1984) *Managing construction projects – a guide to processes and procedures*, Geneva: International Labor Office.

Chandler, A.D. (1966) *Strategy and Structure*, New York: Anchor Books.

Cherns, A.B. and Bryant, D.T. (1984) 'Studying the client's role in construction management', *Construction Management and Economics* 2: 177–84.

Curtis, B., Ward, S. and Chapman, C. (1991) *Roles, responsibilities and risks in management contracting*, special publication 81, Construction Industry Research and Information Association, London.

Davidson, C.H. and Tarek, A.M. (eds) (1997) *Procurement – a key to Innovation*, Montreal: IF Research Corporation.

Franks, J. (1991) *Building contract administration and practice*, London: Batsford.

Franks, J. (1996) *Building procurement systems*, 2nd edn, Englemere, Ascot: The Chartered Institute of Building.

Higgin, G., Jessop, N. Bryant, D.T., Luckman, J. and Stinger, J. (1966) *Interdependence and uncertainty*, London: Tavistock Publications.

Ireland, V. (1983) *The role of managerial actions in the cost time and quality performance of high-rise commercial building projects*, PhD thesis, University of Sydney, Sydney.

Ireland, V. (1984) 'Virtually meaningless distinctions between nominally different procurement forms', in *CIB W65 Proceedings of the 4th International*

Symposium on Organisation and Management of Construction, vol. 1, University of Waterloo, Waterloo, Ontario, 203–12.

Kumaraswamy, M.M. and Dissanayaka, S.M. (1997) 'Synergising construction research with industry development', *First International Conference on Construction Industry Development*, vol. 1, National University of Singapore, Singapore, December 1, 182–9.

Latham, M. (1994) *Constructing the team.* Joint review of procurement and contractual agreements in the United Kingdom construction industry: final report, London: The Stationery Office.

Liu, A.M.M. (1995) *Analysis of organizational structures in building projects*, PhD thesis, Department of Surveying, The University of Hong Kong, Hong Kong.

Masterman, J.W.E. (1992) *An introduction to building procurement systems*, London: E. & FN Spon.

Moss, A. (1994) 'The HKCEC: an unusual and highly successful procurement example', in S.M. Rowlinson (ed.) *East meets West, Proceedings of CIB W92 Procurement Systems Symposium*, Department of Surveying, The University of Hong Kong, Hong Kong, December, 213–20.

MPBW (1970) *The building process – a case study from Marks & Spencer Ltd.*, Ministry of Public Building and Works, Research and Development Bulletin.

NEDO (1983) *Faster Building for Industry*, National Economic Development Office, London: The Stationery Office.

NEDO (1985) *Thinking about building. A successful business customer's guide to using the construction industry*, London: National Economic Development Office.

Newcombe, R. (1994) 'Procurement paths – a power paradigm', in S.M. Rowlinson (ed.) *East meets West, Proceedings of CIB W92 Procurement Systems Symposium*, Department of Surveying, The University of Hong Kong, Hong Kong, December, 243–50.

Rowlinson, S. (1987) 'Design build', Chartered Institute of Building (UK) Occasional Paper, Design Build – Its Development and Present Status.

Sidwell, A.C. (1984) 'The measurement of success of various organisational forms for construction projects', in *CIB W65 Proceedings of the 4th International Symposium on Organisation and Management of Construction*, University of Waterloo, Waterloo, Ontario, 283–90.

Skitmore, R.M. and Marsden, D.E. (1988) 'Which procurement system? Towards a universal procurement selection technique', *Construction Management and Economics* 6: 71–89.

Turner, A.E. (1990) *Building procurement*, London: Macmillan Education.

Veld, J. and Peeters, W.A. (1989) 'Keeping large projects under control: the importance of contract type selection', in *Project management*, 7 (3): 155–62.

Walker, A. (1996) *Project management in construction*, 3rd edn, Oxford: Basil Blackwell.

Walker, A. and Kalinowski, M. (1994) 'An anatomy of a Hong Kong project – organisation, environment and leadership', *Construction Management and Economics* 12: 191–202.

Walker, A. and Newcombe, R. (1999) 'The positive use of power to facilitate the completion of a major construction project: a case study', *Construction Management and Economics*, forthcoming.

Walker, D.H.T. (1994). 'An investigation into factors that determine building construction time performance', Department of Building and Construction Economics, RMIT University, Melbourne.

Wood, K.B. (1975) *The public client and the construction industries*, London: National Economic Development Office.

Part II

Organisational issues in procurement systems

3 Organising the project procurement process

Stuart D. Green and Dennis Lenard

Introduction

This chapter comprises four main sections. First, the client organisation and the importance of strategic briefing are considered. It is the client who is the ultimate customer of the procurement process. The success of every construction project is dependent upon the ability of construction professionals to understand the needs of their client. Despite this it is the interface between the client organisation and the construction team that remains the most problematic. Many construction professionals remain naive about the client organisations for which they work. If clients are to achieve value for money from the construction industry, then construction professionals must learn to manage the software as well as the hardware. As we approach and start the next millennium, to be a good builder is no longer enough.

Second, the concept of group decision support (GDS) is described and the way in which it can be applied to improve strategic briefing. The link is made between GDS and the established practice of using value management to improve communication and understanding during the early stages of briefing. The concept of a methodology is initially defined, followed by a discussion of the ways in which value management differs from the traditional cost emphasis offered by value engineering. Two new methodologies of GDS are then outlined which recent research suggests will be of particular benefit for the purposes of strategic briefing.

Third, the argument that construction procurement cannot be understood in isolation is developed. If progress is to be made, it is necessary to view procurement as a constituent part of an integrated process that cuts across the established disciplines of the construction industry. The argument is developed by drawing insights from a recent case study of a manufacturing facility in Sydney. Particular observations are made regarding the impact of project complexity and the importance of an integrated team approach. Insights are also provided into the design process and the uniqueness of the procurement method.

Last, the issue of serial contracting is addressed. One of the long-standing problems of the construction industry is the need for the project team to climb the learning curve on every project. The team will be faced with many problems that will require innovative solutions, but the lessons learned are rarely carried forward to the next project. It is this discontinuity in process that is one of the major barriers to improved practice in the construction industry. Recent developments in the area of partnering, strategic alliances and serial contracting are at long last making progress. An important underlying influence is the Japanese management practices which have been imported from the manufacturing sector.

Client organisations and strategic briefing

The most problematic areas in the procurement process lie at the interfaces between organisations at different stages of the construction supply chain. Clear communication and clarity across each interface are vital in the establishment of a continuous value stream throughout the process from component supplier to the client organisation. The latest thinking on lean production provides a focus for ensuring that waste is eliminated at each interface. Nevertheless, the interface that remains the most problematic, and the least managed, is that which exists between the client's organisation and the project. The importance of this interface is reflected in the increased emphasis given to strategic briefing. The quest for downstream efficiencies in the supply chain must not be allowed to distract from the need to ensure that construction projects are assessed properly in terms of their contribution to the client's business processes.

Strategic briefing and client business processes

The briefing process can be divided into two major stages, the first of which comprises a strategic review of the client's organisational needs. The second stage is more tactical in nature and is primarily concerned with issues of performance specification. Many corporate clients often perform the first stage before engaging the design team. Prior to becoming involved in issues of design, the need for a new building should be determined by the extent to which it contributes to the client's value chain. The first stage of briefing therefore depends less on an ability to conceptualise built solutions and more on an ability to understand the client's business processes.

The recognition of the initial need to understand a client's business processes offers a promising step forward in that it clearly goes beyond the simplistic checklists offered by the more traditional briefing literature. The emphasis given to understanding the client's strategic context directly addresses the recurring criticism of practitioners that they tend to give far too little time and attention to exploring the problem. It is also persuasive to link client briefing with the latest thinking on business processes. To argue the case for a new facility in terms of business process improvement represents a significantly new approach to formulating the statement of need. However, the skills and knowledge required for strategic briefing are significantly different from those required for downstream management activities. Of paramount importance is the need to understand the complexities of the client organisation.

Understanding organisations

An increasingly influential approach to understanding complex organisations is based on the concept of organisational metaphors (Morgan 1986). Practising managers make use of metaphors every day in order to make sense of their work environment. The various management theories can also be analysed in terms of their underlying metaphors. For example, traditional management theories are characterised by their allegiance to the machine metaphor. They assume that organisations can be engineered in the same way as machines. Organisations are seen to be goal-seeking black boxes which consist of reproducible and interchangeable parts. People are equated with mindless cogwheels. Every employee has a precise job description and a defined limit of responsibility. Instructions flow from the top down and problems flow from the bottom up. Alternative management philosophies are underpinned by different metaphors. Morgan (1986) identified eight metaphors in all, including the organic view, which sees organisations as organisms that need to adapt to different environments. Other influential metaphors are the political view, the cultural view and the cybernetic view. Some metaphors provide conflicting views of organisations. For example, the political metaphor stresses the role of political action whereas the culture metaphor emphasises the need for strong norms of behaviour. It follows that there can never be any single objective interpretation of an organisation.

The concept of organisational metaphors is useful in two ways. First, different insights can be gained by viewing organisations from different metaphorical perspectives. Second, to understand an organisation it is necessary to understand the metaphors that the organisation's managers use to understand themselves. The reality of an organisation will also be influenced by the dominant management philosophy. For example, if management views the company in terms of the simplistic machine metaphor, then the organisation will become progressively more machine-like. Employees who do not like being treated as mindless components will simply leave and be replaced by others. In a similar way, a tendency to view the organisation as a political arena would result in an increase in political behaviour. Different interest groups will be associated with different ideas. Some ideas will be in the ascendancy, others will be in the descendancy. Successful organisations are always characterised by a continuing discussion of how they can improve. Management fads certainly play an important role in this process. TQM (total quality management) gives way to BPR (business process re-engineering) which in turn gives way to next year's big idea. All of these invoke different metaphors and each focuses on different aspects of a company's business. Modern organisations are continually seeking to reinvent themselves and therefore cannot be understood without also understanding the debate between competing management ideas. The danger for modern organisations is that they move from one poorly defined management philosophy to another without ever supporting the rhetoric with rigorous methodologies.

Metaphors and client briefing

The relevance of organisational metaphors to the briefing process has been previously demonstrated (Green 1996). Three characterisations of client briefing were presented and each was analysed in terms of its underlying metaphors. The first characterisation is based on the approach adopted by large bureaucratic organisations such as the Civil Service and the Ministry of Defence in the UK. These types of client tend to be dominated by the machine view. One person is placed in charge of the entire process, and consultation with the building users is seen as a needless digression. The assumption is that the organisation is unitary in nature and that the project objectives can be predetermined in isolation from the aspirations of the building users.

The second characterisation of briefing is based on the view that the activities of briefing and designing are iterative and interdependent. Many client organisations comprise multiple stakeholders, and the existence of a predetermined objective cannot be taken for granted. Such clients are often unable to pre-articulate their objectives until they have engaged with the design process. Client stakeholders and designers collectively learn their way to a shared understanding of the project requirements. This approach also recognises that clients exist in dynamic environments. This means that their requirements often change during the course of design and construction. The layered approach to the briefing process, whereby each decision is delayed until the last possible moment, is consistent with this perspective. The underlying metaphors are more sophisticated than that of the machine view. First, the emphasis on iterative learning between designers and users is suggestive of the cybernetic metaphor. Second, the recognition of different interest groups, each with a valid point of view, reflects the political metaphor. Third, the need to respond to changes in the broader environment is indicative of the organic metaphor.

The third characterisation of briefing is based on the approach developed by UK speculative developer clients in the 1980s. These clients were able to pre-articulate their requirements and were also unitary in nature. They sought to develop standard briefs on the basis of their previous experience. Although designers on individual projects were encouraged to offer ideas on how the brief could be improved, the clients were adamant that they did not want too many ideas. The philosophy was one of responsible innovation. This approach to briefing is still very much in evidence in the 1990s amongst regular clients such as large retail chains. The speculative clients of the 1980s were also notable for their determination to change the outdated practices of the UK construction industry. The implication was that the industry was trapped into fixed ways of working. The underlying metaphor is that of the psychic prison.

It would be a mistake to suggest that any one of the above approaches to client briefing is better than any other. Different approaches to client briefing are necessary for different situations. Although the complexity of the project is certainly an important variable, the adopted approach also depends upon the extent to which the objectives can be predetermined. In situations characterised by ambiguity and complexity any attempt to implement a machine-like approach to client briefing would be doomed to failure. Within the context of modern multifaceted client organisations, such situations are the norm rather than the exception. Clear objectives and a definable product are now the exception rather than the rule. Client organisations are increasingly pluralistic, with different interest groups pursuing different agendas. Objectives are no longer dictated from the top but are generated across the organisation by diverse groups of stakeholders. The stakeholders must be involved in the process. The success of the end project will depend not only upon efficient construction but also upon the development manager's ability to develop and maintain a political constituency for the project brief. Value for money therefore depends not only on managing the hardware of construction but also on managing the software of the client organisation.

Group decision support for client briefing

The achievement of value for money is one of the essential objectives of procurement policy. An emerging understanding has been noticeable in recent years that value for money is not the same as cost effectiveness. It does not matter how cost effective the industry is in design and construction if the end product does not add value to the client's business. The purpose of this section is to investigate the contribution offered by the emerging methodologies of group decision support (GDS). The concept of GDS is in many ways synonymous with that of value management. In the past, value management has often been perceived by the academic community to be something of a black art. However, the guiding framework of GDS provides a rigorous justification which has previously been absent. There are several methodologies of GDS which potentially offer a significant contribution to strategic briefing. In some cases the label of value management will be useful, in others it may be appropriate to use alternative labels. The emphasis on the word *methodology* is important. Value management practitioners tend too often to focus solely on techniques which are largely meaningless in isolation from any guiding principles. In the following subsections the nature of methodology is defined and then the methodological distinction between value management and value engineering is discussed. The concept of GDS is introduced and linked to strategic procurement problems. It is further argued that the application of traditional value engineering is limited to narrow technical problems. It is assumed that readers are already familiar with SMART value management as developed in accordance with the principles of GDS (Green 1994). The latest developments are described which focus on the use of two other GDS methodologies.

The nature of methodology

It is initially necessary to define the term methodology and the way in which it differs from a method or a technique. The notion of a methodology is much broader than that of a method in that it includes a set of guiding principles which are based on an underlying theoretical position. This interpretation is supported by Checkland

> the essence of a *methodology* – as opposed to a method, or technique [is that] . . . it offers a set of guidelines or principles which in any specific instance can be tailored both to the characteristics of the situation to which it is applied and to the people using the approach.
>
> *(idem* 1989b: 101, emphasis in original)

It is therefore apparent that most descriptions of value management (for example, Miles 1972; Dell'Isola 1982; Kelly and Male 1993; Norton and McElligott 1995; Parker 1985) are theoretically immature in that they tend to concentrate on prescribing methods and techniques rather than issues of methodology. The commonly accepted method of value management is that of the job plan. Specific techniques include function analysis, brainstorming and life-cycle costing. Unfortunately, the philosophical and theoretical beliefs which lie behind the advocated methods and techniques are rarely made explicit. If the different methodologies of value management are to be understood it therefore becomes necessary to infer the often taken-for-granted underlying theoretical positions.

Value engineering compared with value management

Previous work has articulated two alternative methodologies of value management which are fundamentally different in terms of their underlying theoretical beliefs (Green 1994). The first methodology borrows its philosophy from the scientific method and assumes that problems are essentially technical in nature and that they exist independently of human perception. The second draws from the philosophy of social science and emphasises that differing perceptions are an essential ingredient of any real-world problem. For the purposes of enunciation it is convenient to label the former as value engineering and the latter as value management.

Value engineering

If the proposed distinction between value management (VM) and value engineering (VE) is to be meaningful in practice, the two methodologies must be seen to differ, not only in terms of theory but also in terms of their means of implementation. VE, as epitomised by Dell'Isola (1982), is invariably implemented retrospectively in response to a projected cost over-spend. A review of the construction case studies described in the past five annual conferences of the Society of American Value Engineers (SAVE International) reveals this to be the dominant interpretation in the USA. It is implicitly taken for granted that design solutions form part of a reality which exists out there. The identification of the optimal design solution is therefore simply dependent upon exploring the solution space. This posi-tion is epitomised by the way in which descriptions of function analysis assume that the underlying real function can be revealed by the application of a prescribed technique. The overriding assumption is that the problem exists independently of the perceptions of the project stakeholders. Whilst this approach may well be justified if value engineering is applied to a build-ing component, it is rarely justified for the soft, messy and ill-defined problems which often dominate during the early stages of building design.

Value management

In contrast to the tradition of VE, the broad theoretical framework for SMART value management is provided by the concept of GDS, which can be defined as 'any designed process that supports a group of people seeking individually to make sense of, and collectively act in a situation in which they have power' (Bryant 1993: 19).

The concept of GDS represents a decisive break with the US tradition by rejecting both the optimising paradigm of operational research and the associated terminology of function analysis. The above definition presents an obvious analogy to the role of VM in aiding design decision-making in general and the briefing process in particular. It should also be recognised that there is a significant difference between the provision of GDS and the narrower concept of decision support to an individual. GDS differs in that it places less emphasis on substantive data and more emphasis on consensus building and the decision-making process. Given that building design is invariably a group activity which includes designers and client representa-tives, it is clearly GDS which is relevant. This is especially true for multi-faceted client organisations where different interest groups possess conflicting objectives. Indeed, once the intellectual baggage of optimisation is jettisoned, the concept of VM becomes almost synonymous with that of GDS. The above quoted definition of GDS could therefore be suggested as a meta-definition of VM.

From the GDS perspective, VM is primarily concerned with resolving ambiguity by constructing a shared consensus of the project objectives. The process of sharing different perspectives is facilitated by the use of decision-structuring techniques. The function of a building is therefore perceived to be something which is socially negotiated rather than revealed. There is no pretence of identifying optimal solutions; the emphasis lies on the process of VM and the extent to which participants learn from their involvement.

Limitations of existing approaches

Notwithstanding the established success of SMART value management, it has become increasingly clear through sustained action research and consultancy that the methodology is by no means generally applicable. Whilst the overt rationality of decision conferencing can bring confidence to some clients, there are others who feel uncomfortable with its reliance upon quantitative decision modelling. The experience of implementing SMART value management within large client organisations has also demonstrated that it becomes less effective over time. Once participants become familiar with the weighting procedures they become increasingly proficient at manipulating the numbers to ensure the selection of their previously favoured option. VM is only of benefit if it stimulates people to innovate and think critically about their decisions and procedures. If VM becomes a standardised routine, implemented without active participation and commitment, it will not achieve its objectives. It is therefore considered important that VM practice is not limited to any single methodology.

Group decision support

The potential application of the existing methodologies of GDS to the unstructured problems of early building design is readily apparent from the comments of Rosenhead (1989: 1), who stated that 'distinctive features of these novel approaches include an aim of partial structuring of previously unstructured situations (rather than the solution of well-structured problems), and a process involving participation as a key component'.

It is important to recognise that such approaches make no attempt to identify optimal answers. The emphasis lies on constructing a social consensus regarding the nature of the problem together with an agreed course of action. It is particularly important to secure the involvement and commitment of the problem stakeholders. It is also recognised that the very act of modelling will inevitably alter the nature of the problem.

The following two GDS methodologies are considered to be particularly deserving of further research within the context of strategic briefing:

- soft systems methodology (Checkland 1981, 1989a);
- strategic choice (Friend and Hickling 1987).

Both methodologies recognise the importance of a skilled facilitator and the interdependence between decision content and decision process. In contrast to hard operational research, the various models upon which the methodologies are based are seen to be facilitative devices rather than representations of reality.

Soft systems methodology

Within the constraints of this chapter, it is not feasible to describe soft systems methodology (SSM) in any detail. A full appreciation of its richness can be achieved only by reference to the source literature. Nevertheless, the book would not be complete without offering a brief description. SSM has been specifically developed in response to dynamic, multiperspective social problems which defy any attempt at solution. Of particular significance within SSM is the conceptual distinction between the real world and systems thinking about the real world. The real world is seen to generate problems; systems thinking is used to produce conceptual models which are then compared with the real world in order to provide meaningful insights. In theory, SSM provides a never-ending process of enquiry which terminates only when it ceases to provide meaningful insights. SSM is implemented as a participative process where a facilitator works with the problem stakeholders. In contrast to VM, the methodology is likely to unfold over a series of workshops ranging in length from two hours to one day. SSM provides a designed methodology which enables the facilitator to make explicit different interpretations of the problem based on different *Weltanschauungen*, or world-views. Conceptual systems models are then built in accordance with each of the identified *Weltanschauungen* and different insights are derived by comparing each of these models with reality. The recognition that human activity systems are subject to different interpretations from different points of view, each of which are equally valid, is in harsh contradiction to traditional VE. Whereas function analysis seeks to produce models of reality, SSM seeks only to produce models which are relevant to the debate about reality. Such is the distinction between operational research and GDS.

   The increasing concern amongst clients for construction professionals to understand their business processes before embarking on design makes SSM especially applicable in the current context. SSM offers a means by which construction professionals and client representatives can derive a common understanding of the client organisation's business processes. A further source of concern relates to the terminology of SSM, which undoubtedly provides something of a barrier to the uninitiated. Unfortunately for those who have an in-built aversion to jargon, the methodology of SSM is largely inseparable from its terminology. It is simply not possible to appreciate the significance of SSM without understanding the associated language of systems thinking together with the guiding framework of GDS. However, in practice there would be no need to expose the participating client representatives to the terminology of SSM, nor would it be necessary to explain the methodology to them in advance. All that would be required would be a commitment to a series of participative briefing workshops, which could vary in length from as little as two hours to a maximum of one day, as already stated. The first such session could be introduced as one of problem definition and the rest of the methodology could then be introduced one step at a time. If the client organisation already uses the language of BPR, then SSM could be easily presented as an approach to process modelling. (For further information, see Green and Simister 1999.) Of course, some leading VM practitioners are already including BPR amongst their list of services. However, unlike SSM, existing approaches to BPR tend to be strong on rhetoric and weak on rigour.

Strategic choice

Strategic choice is rooted in the socio-technical approach pioneered by the Tavistock Institute in London during the 1970s. The approach is empirically based in that it draws from studies of strategic decision-making in practice. Once again, it is facilitator-driven with no specific constraints regarding the number or length of workshops. Strategic choice is framed around four complementary modes of decision-making activity, between which the decision-makers are likely to iterate.

The first mode, described as the shaping mode, is concerned with problem formulation. Key techniques include the graphical identification of, and linkage between, decision areas. This enables the decision-makers to identify the most urgent problems and agree upon an initial problem shape.

The second mode is labelled the designing mode, during which the facilitator steers the participants towards the identification of different options. Of particular importance is the grouping of different combinations of options into discreet decision schemes. It is recognised that whilst some options would be compatible, others would be mutually exclusive.

The third mode is the comparing mode and consists of a sequence of techniques which seek to compare the benefits of alternative decision schemes. These techniques differ from those of decision analysis in that they allow for a combination of quantitative and qualitative comparison. As such they can readily be absorbed into the framework of VM, thereby avoiding any dependence upon a complete set of numerical utility ratings. Of further importance is the way in which the comparing mode conceptualises three different types of uncertainty:

- uncertainties pertaining to guiding values (UV);
- uncertainties pertaining to related decision fields (UR);
- uncertainties pertaining to the working environment (UE).

UV relates to lack of clarity regarding objectives, that is, uncertainty of a political nature. The resolution of UV lies in consensus-building exercises such as SMART value management.

UR relates to the interconnectiveness between decision areas, that is, the decision-makers are unclear on the effect on other decision areas. The response here may be to reframe the decision area or to consult with others beyond the immediate constituency of problem-owners. Existing VM practice tends not to conceptualise UR as a distinct area of uncertainty.

UE is the kind of uncertainty which is normally dealt with through risk management. UE is essentially technical in nature and can be reduced by further research by means of surveys, forecasting exercises or cost estimations. This uncertainty is often dealt with in traditional VE by means of the inclusion of risk analysis.

The final stage of strategic choice is described as the choosing mode and is concerned both with making immediate decisions and with devising a strategy for managing those decisions which are best made in the light of further information. The outcome of any particular meeting would therefore always include immediate commitments to action and also strategies for resolving identified areas of uncertainty to aid future decisions. The latter aspect has some commonality with the established practice of maintaining risk registers.

Perhaps the most important aspect of strategic choice is the way in which it provides a conceptual framework which embraces not only VM but also risk management. Indeed, it can be argued that it is meaningless to try to assess value independently of risk, or vice versa. It is no coincidence that many existing facilitators offer both services. No VM exercise can afford to ignore risk; the only issue is whether or not risk is made explicit. Whilst further research in this area is ongoing, it is already apparent that risk management and VM can no longer be considered to be two separate entities.

Integrated facilities, design and management

The need for construction professions to understand the complexities of client organisations is partly to account for the trend towards longer-term relationships. Within this frame of reference, the activities of facilities management, facilities design and construction management become constituent parts of an ongoing seamless process. This wider vision of the procurement process requires much greater integration between the various construction specialisms. Indeed, the changing pattern of relationships raises the question of the need for a radical rethink on the way the industry has been traditionally subdivided. The principles are best illustrated by the use of case studies.

Integrated construction: client's briefing process

In 1987 the ING Bank (formerly the NMB) of the Netherlands moved into its newly completed headquarters buildings in Amsterdam. The new buildings were unique in many ways and their success in a number of areas has provided a fine example of the ways in which cross-disciplinary design teams working towards an integrated design solution from the commencement of a project can produce a wide range of benefits for building owners and occupiers. If passive or low-energy systems are utilised then environmental benefits can flow on to the community at large.

The bank's brief was simple and open-ended: it was to provide a facility which would:

- integrate art, natural materials, sunlight, green plants, energy conservation, low noise levels and water;
- be functional, efficient and flexible;
- be human in scale;
- have low running costs.

The buildings were designed by an integrated team consisting of architects, a construction engineer, a landscape architect, an energy expert (a physicist) and artists, as well as the bank's own project manager. All of the consultants on the team were involved from the project's inception, and all consultants were free to comment on any aspect of the design. The results are exemplary: energy consumption is less than one-tenth of that of the bank's previous headquarters, with attendant savings in running costs of around $US 2.4 million per annum, absenteeism was reduced by around 25 per cent, and the bank's corporate image was enhanced to the extent that it moved from fourth to second ranking amongst Dutch banks. Careful integration of the building envelope, services and extensive daylighting has led to significant reductions in greenhouse emissions associated with the operation of the building (Romm and Browning 1994, 1995).

The most important lesson which can be learnt from the ING project is the importance of the multidisciplinary team approach to design if truly better buildings are to be realised. Integrated design and construction is much more than mere coordination. It requires a close working relationship between all members of the team and, importantly, all team members, regardless of their discipline, are on an equal footing within the team. For integrated design and construction strategies to work, several components must be present. Berry (1995: 29) suggests that 'clear thinking and a shared vision are perhaps the two most important requirements for the individual and the team'. However, if the process is to be successful it requires that the client be both flexible and informed. One of the primary reasons for the success of projects such as the ING Bank was the large degree of client involvement throughout the design process. The initial, conceptual brief outlined above was very broad in its aims; however, the design team worked closely with the bank's representative as the details of the brief and the design were developed.

Integrated construction: refurbishment case study

The project described here involved the refurbishment of a five-level process manufacturing facility at Botany, an inner suburb of Sydney, undertaken on a design and construct basis. The facility was operational for the duration of the project. The project had many notable features.

- Construction occurred around the working operating plant which presented a number of programming, productivity and occupational health and safety challenges.
- The client let the contract for design development at a lump sum price. After submitting an initial project estimate, the contractor accepted the full commission to provide detailed design and construction. This commission involved an intensive 18 month programme of investigation, construction planning and costing prior to any work commencing on site.
- Planning was a key part of the project's success. The building programme was coordinated with the operating plant's maintenance schedules so that full advantage could be taken of operational down time.
- There was a partnering arrangement between the owner and the contractor, with no intermediaries involved. The partnering plan related to the actual development process and stringent occupational health and safety objectives. The parties shared their knowledge and experience and felt they had benefited from the joint development, documentation and training of staff in these areas. A full exchange of company and project aims contributed to the success of the project.
- Workplace reform was targeted as a way of improving productivity. The contractor placed a strong emphasis on the workplace being a learning environment, and staff and subcontractors assuming more responsibility. The contractor's goals included breaking down demarcation, multi-skilling workers, career development and counselling, increasing worker responsibility via minor works cost sheets and assisting subcontractors to implement reforms.

Project complexity

The process plant provided packaged food products for the Australian and the Asia Pacific region markets. The continuation of production within the client's planned schedule was a key objective in this project. This created an extremely high level of complexity since the facility had to be rebuilt around working mechanical equipment ensuring that production, safety and hygiene were all maintained in accordance with strict guidelines.

The original building was constructed in the 1940s of steel frame, timber floors (five levels) and a brick facade. Over the years that the building has been used, the process plant contained within it has been changed many times. The original timber floors of the building had been overlaid with concrete which had subsequently been modified structurally by penetrations required for new or modified process plant. All of these modifications, together with substantial corrosion to many of the original steel beams, had caused the building to become structurally unstable. Also, the building, because of its age, did not comply with fire safety regulations.

Design and construction was the contract strategy method chosen, with full project management services supplied by the construction contractor. The contract period for stage one was 120 weeks with a target schedule of 106 weeks, and at the time the case study was undertaken the project was approximately 40 per cent complete and on schedule.

Budget constraints and production requirements demanded that the equipment could not be shut down for any delays other than for scheduled maintenance. The production equipment within the facility had been recently upgraded and specially engineered to fit within the confines of the building.

The major tasks involved in the rebuilding of the structure were:

- addition of temporary external bracing for the duration of the project;
- demolition and reconstruction of the floors;
- addition of a perimeter concrete beam at each level to distribute lateral loads;
- addition of a new structural service core and stair;
- improvement of the bearing capacity of footing piers;
- concrete encasement and fireproofing of steel columns;
- removal and replacement of corroded steel columns and beams;
- demolition of the existing masonry facade and replacement with a lightweight metal sandwich panel system.

The main constraint was that the refurbishment work had to be done while the production equipment and food processing plant was fully operational. This meant that work areas had to be sealed off from production areas to maintain cleanliness and hygiene. This also restricted the materials which could be used in the project. For example, no timber was allowed because of the possibility of undetectable splinters falling into processing equipment.

One example of the problems encountered was the need to suspend machinery from the floor above while the floor was demolished, formed up and poured. Throughout this work the machinery had to remain operable. Once the floor was strong enough the machinery could be relocated on the floor slab.

A second example involved encasing one of the columns with concrete. This particular column had an equipment control panel attached to it with hundreds of electrical circuits connected to it from various parts of the building. This panel and its wiring had to be carefully labelled, disconnected, moved and reconnected, just to encase the column with concrete. Man-handling the materials, including heavy steel beams, was another significant difficulty encountered.

An integrated team approach

The client, a major multinational food producer, had a number of specialist production personnel interfacing with the contractor's site staff to ensure that the client's needs with respect to safety, cost and continuous plant operation were communicated and achieved. The on-site relationship between the client's representatives and the construction team was excellent and a definite feeling of 'partnering' emerged.

Overall project time, in terms of speed of construction, was of less concern to the client if food processing continued throughout the rebuilding process. However, construction time was a major consideration to the contractor. The contractor had been negotiating for up to five years with the client with a view to obtaining construction work from them. The contractor, in recent years, has developed the expertise to undertake the complete design, construction and commissioning of process plant facilities.

This project required significant interaction between the client and the contractor in many areas of the redevelopment. Both parties shared their knowledge and learned from the other. As an example, many of the client's standards for building construction relating to (internal) health and hygiene standards were outdated and were significantly revised after discussion and consultation with the contractor. The other major area in which this partnership has had tangible benefits for both parties was in the formulation and documentation of training in safety and quality standards for staff of both organisations.

The architect was appointed by the project manager/contractor who had worked previously on other refurbishment projects. The architect was responsible for the design coordination and documentation of the building works. The structural engineers had a difficult task in redesigning the structure for the building while giving consideration to the constructability issues made complex by the occupancy of the client and operating machinery. The other problems were mainly those of the unknown, such as the bearing capacity of the foundations and unexpected deterioration of existing structural elements discovered during demolition.

The contractor only appointed experienced subcontractors with whom it had worked on previous projects. The need for safety and hygiene were critical, especially during stages of demolition which also included asbestos removal. In all there were 18 separate subcontractors involved in structural work on the project and six subcontractors involved in the finishes. The key subcontractors were demolition, structural steel, form work, reinforcement and concrete, the various mechanical services and finishes. All of these key subcontractors were incorporated into the integrated team.

Design process

The extended design and planning phase of the project gave the team the opportunity to consider many alternatives for construction methods and some opportunity for the use of new technology. These choices were usually reviewed by the project manager and the client's representatives. Unsuitable alternatives were eliminated and the remaining ones evaluated on a 'constructability and cost' basis. Cost control during the design stage was very important and was the responsibility of the project manager. Owing to the fact that the client had limited funding approval for the project and wanted a lump-sum price, the project pricing had to be extremely accurate and within the client's budget. The cash flow of the client also had to be taken into consideration during the construction planning so that progress claims would be made at times agreed with the client to suit cash-flow projections.

The project organisation of the construction team for this project was, for the most part, based on organisation principles which the company commonly used. However, the site manager pointed out that most projects were unique in some way. The major factor which influenced the decisions regarding the formation of this organisational structure was the structure of the client's team. The site manager wanted to ensure that each specialist person on the client team had a specialist person to interface with on the construction team. The rest of the construction team was determined by the needs of the project. This approach was adopted to ensure that every possible opportunity was available for communication between the contractor and the client, especially because of the heavy client involvement in the project. Communication on such a complex project was essential for the contractor both to ensure the success of the project and to build the relationship that would last for the duration of the project and for future projects.

The project team was divided into sections for services and for structures, which allowed for a relatively flat structure. This structure has not changed over the life of the project thus far, although a few of the people in certain positions have changed. The project manager made certain that the people on this project were the best available and had extensive experience in refurbishment.

Uniqueness of the procurement method

As mentioned previously, this project was designed and planned simultaneously. Because of the need to rebuild so much of the structure, including floors under machinery, there was no other way to proceed with this project than to have an extensive period of on-site investigation, design, documentation, planning and pricing. The client granted a preliminary works commission worth over US$2 million to the project manager. This commission included the documentation of existing services in the building together with the preliminary project design and the preparation of a lump-sum price for the new works. Full-time construction planners for the structure and services were costed into the commission. This allowed the contractor to give consideration to alternative methods and to make sure that the solutions were constructable. The design team prepared numerous method sheets outlining step-by-step instructions for almost every task on the project. These method sheets were continuously updated and added to as the project proceeded.

Serial contracting: the key to innovation

Serial contracting acknowledges the need to establish long-term multi-disciplinary teams and the important role that the client plays in promoting innovation and idea generation in the design and construction phase. More importantly, though, the ideas must be passed on from project to project.

Idea generation through effective partnership and linkages in the project team

The importance of the project team has been examined in detail (Lenard, for CII 1996b). That examination indicated that historically, project teams (because of the single-project focus) were seen as a barrier to, rather than a generator of, innovation in the construction procurement process. It was also indicated that the concept of long-term alliances may be an effective means of overcoming this barrier particularly when the life cycle of the alliance spans many projects. The current literature on partnering (strategic alliances) in the construction context emphasises the need to establish common goals between the parties for the intermediate to long term. In fact, a survey conducted for the Construction Industry Institute (CII) in 1996 indicated that establishing common goals was critical to the success of such arrangements (Lenard, for CII 1996a).

The establishment of common goals for a period greater than a single project can often foster new approaches and solutions to project problems that lead to exponential improvement. The use of mission statements and the encouragement of entrepreneurial behaviour can be useful in creating the right innovative environment. However, the existing environment is limited in promoting the concept of sharing values and goals because of the current stratification and the linear single point accountability that predominates in traditional lump-sum tendering.

The role of the project team could also be enhanced if project staff were drawn from a range of backgrounds, covering functional specialities and creativity from marketing to project technical capability. Where possible, research and development specialists from the individual companies should also participate. Ideally, the project team should have access to interactive information systems incorporating costing, design and construction solutions for previously completed projects.

The Japanese *keiretsu* (the association of a number of industries centred on a bank) has proved to be particularly effective in reducing time to market and creating long-term competitiveness. Variants of the *keiretsu* are emerging in North America and Europe, with companies such as Ford and IBM acquiring equity positions in suppliers and participating in various external organisations. Fundamental to the success of such partnerships is a rapid and free information exchange and major business transformations where entire processes (both production and management) are integrated. Paper-based systems, bureaucratic approval processes, labour-intensive clerical activities, project development phases, production cycles and multi-layered decision-making processes are being replaced by source data capture, integrated transaction processing, electronic commerce, real-time systems, on-line decision support, document management systems and expert systems. These tools emerge as mechanisms to drive change and create a climate for innovations.

The importance of serial contracting

Serial contracting is underpinned by a strategic alliance that promotes a relationship between a client and a contractor that is for the intermediate to longer term and spans many projects. The alliance is a voluntary, cooperative arrangement between stakeholders in the procurement of constructed facilities aimed at maximising the efficient use of resources towards achieving identified goals and minimising conflict. The longer-term alliance complements formal contractual relationships by defining the interpersonal relationships most conducive to realising those responsibilities.

The motivation for serial contracting is quite straightforward and in many ways reflects key failings of the construction industry worldwide. In particular, organisations enter longer-term arrangements because of conflict, misunderstandings and poor dispute management in previous construction projects (organised on a one-off basis) that have resulted in litigation, cost blow-outs, delays, wastage, compromised quality and a fear to innovate.

Contracting in the construction industry is a high-risk activity in every sense of the term. Whether it be the relationship with a supplier, subcontractor, employees, clients or whomever, the chances of conflict, misunderstanding and litigation are high. Ultimately, costs blow out, quality is compromised and the opportunities to innovate diminish. Potential commercial partners are encouraged to think and act defensively, to expect and prepare for the worst. This invariably entails a significant allocation of resources in anticipation of potential and actual claims, resulting in a proliferation of contractual specifications and amendments.

> Accompanying the voluminous increase in documentation has been the ever-increasing scrutiny and interpretation of contractual documents, not so much to ascertain the true intent but rather to discover legal uncertainties which may provide an avenue to future redress. Even if a contract free of discrepancies, ambiguities and legal anomalies was somehow achieved, uncertainties would still arise; further, an underlying feature of most projects has been the constant dynamic and turbulent change throughout the project life, the magnitude of which varies directly with the project size.
>
> The industry has recognised that to become competitive and to maintain quality, it is necessary to embrace an alternative to existing practice; a technique which will offer an opportunity to revive the industry and restore confidence in it.
>
> (Turner, 1994: 1)

A strategic alliance resulting in serial contracting is an explicit, voluntary, legally informal cooperative arrangement between two or more parties involved in some common endeavour aimed at maximising the efficient use of resources towards achieving identified goals and minimising conflict for a term greater than a single project.

In the late 1980s the construction industry in the USA realised that it needed to remedy a serious increase in all forms of litigation. The industry quickly recognised the need for a team approach to projects that reinforced a win–win attitude rather than the traditional competitive notion of winners and losers, even among collaborating contractual partners. Strategic alliances were formed and often these resulted in serial contracting.

A major factor in the rapid development of such alliances in the USA was DuPont Engineering's desire to find a new approach to contracting which would help it combat foreign competition. DuPont believed that in order to maximise the competitive advantages flowing from total quality management, continuous improvement and cost reduction, it was necessary to improve communications between project collaborators to specify shared risk more precisely and to identify openly and carefully criteria for measuring success and failure. Fluor Daniel was the first contractor to join DuPont in a longer-term relationship.

Project-based alliances

A longer-term alliance may also include project-based alliances. Project-based alliances were initially formulated by Colonel Charles Cowan of the Portland Division of the US Army Corps of Engineers. In 1992 Cowan reported the results of a large number of partnered projects in the following summary (Cowan 1992):

- no litigation
- joint satisfaction of all shareholders;
- reduced cost growth from an average of 10 per cent to 3.3 per cent;
- improvement savings of 7.1 per cent and 5.2 per cent on two projects;
- completion on time or ahead of schedule;
- reduction of paperwork by approximately two thirds;
- reduction of overhead of projects from 24 per cent to 11 per cent in three years;
- synergism of sharing knowledge and teamwork.

Since 1992 the concept has gained wide acceptance world-wide, especially by governments and manufacturers requiring large numbers of constructed facilities. The participants or partners in a strategic alliance comprise those stakeholders which are directly involved in the delivery of projects. Stakeholders can therefore be regarded as the end user; principal subcontractors; consultants; plant or material suppliers; principal contractors; design consultants; contractors' consultants and so on. The underlying principle is that the arrangement is sustained for a succession of projects.

The elements of a successful long-term alliance

Cowan utilised notions developed in Stephen Covey's book, *The seven habits of highly effective people* (1989). The seven habits are:

- be proactive – the principle of self-awareness, personal vision and responsibility;
- begin with the end in mind – the principle of leadership and vision;
- put first things first – the principle of managing time and priorities around roles and games;
- think win–win – the principle of seeking mutual benefit;
- seek first to understand, then to be understood – the principle of empathetic listening.
- synergise – the principle of creative cooperation;
- sharpen the saw – the principle of continuous improvement.

Cowan restructured Covey's habits as seven key elements:

- commitment,
- trust,
- equity,
- mutual goals and objectives,

- implementation,
- joint process evaluation,
- the dispute resolution process.

Commitment is fundamentally a responsibility of senior management. It must be visible, supportive, ongoing and sensitive to organisational change in relation to continuing partnering skills training and development. At the organisational level commitment is realised in a champion, though, ultimately, the aim of a successful alliance is to move beyond individuals and to emphasise processes.

Trust is the central element of partnering and the hardest to achieve. Beyond honesty, trust is based on consistency, fairness and appropriate and timely behaviour.

Equity reflects a sense of proportionality and balance that transcends simple fairness. Importantly, it implies an understanding of and respect for the interests and needs of stakeholders.

Mutual goals and objectives furnish the framework that will define any particular partnering process. Commonly, such goals include:

- project completions on or before time;
- project completions within budget;
- maintaining desired project quality;
- meeting the financial goals of each participant;
- clarity of objectives through value management;
- value engineering savings and technological innovation;
- adherence to health and safety standards;
- avoidance of litigation;
- project-specific goals.

Implementation entails the establishment of mechanisms for realising and monitoring each party's objectives. From a strategic perspective, this necessitates:

- an understanding of the underlying objectives of the relationship;
- the selection of company participants;
- the evaluation of each party's objectives.

Joint Process Evaluation reflects the need for ongoing monitoring of the relationship, not only to ensure adherence to the agreement but also to learn for future projects.

The issue resolution process provides the vehicle for resolving difficulties and disputes as they arise in a fashion agreeable to all. In particular, there must be a mechanism for dealing with problems at the point at which they arise, escalating to higher levels of management only if necessary.

The key elements of the arrangement are a shared understanding of, and commitment to, common goals; individual, realistic understanding of the expectations, interests and values of other stakeholders in the arrangement; an atmosphere of trust grounded in a commitment to maintaining the long-term relationship; an awareness of the spread of risk and reward; and a willingness to resolve conflict creatively. The realisation of this approach is discussed in detail by Matthews in Chapter 11 of this book.

Projects, organisations, stakeholders and contracts do not exist in a vacuum; they are embedded in a variety of industrial, social, legal and other contexts that determine what is possible and, often, what is desirable. The procedure, then, is a matter of interorganisational and intraorganisational cultural assessment and development. The processes, habits and techniques of organisations, as well as all those indefinable features of group life that give an organisation its identity, constitute its culture, its personality, so to speak.

Likewise, the interaction of organisations reflects industrial and other cultural factors. In establishing a long-term relationship, stakeholders endeavour to shape a new and more profitable culture that will define their interaction. They are explicitly accepting the need to do things better and smarter – and more cooperatively. No arrangement can succeed without a favourable cultural backdrop. No matter how sincere individual chief executive officers or project managers might be, if the systems and personnel of their organisations are unable to reflect the trust, commitment and vision underpinning an arrangement, the agreement will fail. Liu and Fellows provide a detailed discussion of culture in Chapter 6 of this book.

As an agent of oganisational change, then, long-term relationships can be powerful. First, they focus on the need of the organisation to succeed (in the long term) in a potentially hostile external environment. After all, if there were no perceived threats to organisational success, why make more commitments than one is legally obliged to? Second, by addressing specific projects and commercial relationships, the relationship is grounded in reality rather than mere wish fulfilment: arrangements will work or collapse in the real world and with real effects. Third, internal organisational change is driven by interaction with other stakeholders and the need to adapt; the frame of reference for change is established by necessity rather than convenience.

The benefits of long-term strategic alliances

Turner sums up the advantages of such a process by pointing out that

> [for] all project participants, cooperation is a high leverage effort; it requires extra management and time up front, but the advantages accrue in a more harmonious, less confrontational atmosphere and hopefully, completion of a project without litigation and unexpected expense.
>
> (*idem* 1994: 1)

For the constructor, of course, the advantages are similar, as they are for other stakeholders. Cowan (1990, 1992) itemises the potential benefits as follows.

- Benefits to the project owner:

  - reduced exposure to litigation through open communication and issue resolution strategies;
  - lower risk of cost overruns and delays because of better time and cost control over the project;
  - a better quality project, because energies are focused on the ultimate goal and not misdirected to adversarial concerns;
  - potential to expedite the project through efficient implementation of the contract;
  - open communication and unfiltered information allow for more efficient resolution of problems;
  - lower administrative costs because of elimination of defensive case building;
  - increased opportunity for innovation through open communication and elements of trust, especially in the development of value engineering changes and constructability improvements;
  - increased opportunity for a financially successful project because of non-adversarial win–win attitude.

- Benefits to the project contractor:

  - reduced exposure to litigation through communication and issue resolution strategies;
  - increased productivity because of elimination of defensive case building;
  - expedited decision-making with issue resolution strategies;
  - better time and cost control over project;
  - lower risk of cost overruns and delays because of better time and cost control over project.

- Benefits to the project architect or engineer and to the consultants:

  - reduced exposure to litigation through communication and issue resolution strategies;
  - minimised exposure to liability for document deficiencies through early identification of problems, continuous evaluation and cooperative prompt resolution which can minimise cost impact;
  - an enhanced administrative role in the decision-making process, as an active team member in providing interpretation of design intent and solution to problems, as well as a reduced administrative cost

because of elimination of defensive case building and avoidance of claim administration and defence costs;
- increased opportunity for a financially successful project because of non-adversarial win–win attitude.

- Benefits to the project subcontractor and suppliers:

  - reduced exposure to litigation through communication and issue resolution strategies;
  - increased opportunity for innovation and implementation of value engineering in work as a result of the equity of involvement in the project;
  - potential to improve cash flow as a result of fewer disputes and withheld payments;
  - avoidance of costly claims and savings in time and money, as a result of improved decision-making;
  - an enhanced role in the decision-making process as an active team member.

By-products

By addressing the human element in the effort to build a team environment, stakeholders find themselves in a new mode of thinking about and dealing with people. Among the project personnel, and within the stakeholders' own business organisation, work can become more meaningful and fun. Morale and an *esprit de corps* are developed, with a heightened awareness of business and life. A by-product of demonstrating integrity and fair dealing is the respect of others. In the long term, that respect produces a reputation of true value in the industry.

Conclusions

This chapter has addressed several new developments in the management of construction procurement. It was initially suggested that strategic briefing often receives too little time and attention. This is a recurring problem throughout the global construction industry. There is no other industry that has invested so little time and attention in investigating the needs of its clients. The construction industry will not achieve satisfied customers by improving technical effectiveness alone. The delivery of a known product within specified constraints of time, cost and quality contributes only half of the equation. Construction professionals must also be able to manage and maintain a political constituency within the client organisation. If this is to be achieved they must acquire new skills. The control and command paradigm of project management is no longer sufficient. When faced with ambiguous objectives, the emerging methodologies of group decision support provide an important point. To some extent, the same principles are evident within the established discipline of value management. However, the literature on value management has too often in the past fallen short of satisfying the requirements of academic rigour.

A further barrier to enhanced performance in the construction industry is the long-standing fragmentation into separate disciplines. The separation of the construction process into discreet activities performed by different professions owes much to historical precedent and little to effective process design. Whilst it is easy to argue in support of a radical process re-engineering exercise, such initiatives too often ignore the vested interests that oppose any such change. The construction industry is perhaps more resistant than most to change. Innovation therefore tends to be incremental rather than radical. One such example is the trend towards integrated design and construction. This chapter has presented two case studies that provide many lessons for practitioners. The argument has also been made in support of further integration between facilities management and construction procurement. Progress in this respect remains slow. There is no doubt that the reluctance of construction professionals to enhance their skills beyond the construction domain is a significant barrier. Nevertheless, the number of experienced professionals who are seeking to enhance their qualifications with MBA-type qualifications bodes well for the future.

Perhaps the biggest barrier to innovation in the construction industry is that it remains project-based. However, the leading clients and contractors are also acting to establish different ways of working. Concepts such as partnering, serial contracting and strategic alliances are all part of a trend away from approaching construction on a project-by-project basis. Serial contracting is beginning to change the mould of the industry, certainly amongst the leading firms. The argument in support of this trend is persuasive. The industry has much to gain in terms of an enhanced capability for innovation. In addition to the advantages of improved communication and understanding, growth of serial contracting can also reduce the level of uncertainty for both client and contractor. However, there is one message that stands out above all the others. If improvements in procurement are to be gained then an holistic perspective must be adopted. Isolated improvements will not be sustainable unless the big picture is also taken into account.

References

Berry, J. (1995) 'Integrated design', *Building Services* (November): 29.

Bryant, J. (1993) 'Supporting management teams', *OR Insight* 6(3): 19–27.

Checkland, P.(1981) *Systems thinking, systems practice*, Chichester, Sussex: John Wiley

Checkland, P. (1989a) 'Soft systems methodology', in J. Rosenhead (ed.) *Rational analysis for a problematic world*, Chichester, Sussex: John Wiley, 71–100.

Checkland, P. (1989b) 'An application of soft systems methodology' in J. Rosenhead (ed.) *Rational analysis for a problematic world*, Chichester, Sussex: John Wiley, 109–19.

Construction Industry Institute CII (1996a) 'Partnering – models for success', Research report 8, Construction Industry Institute, Adelaide.

CII (1996b) 'Innovation: the key to competitive advantage', Research report 9, Construction Industry Institute, Adelaide.

Covey, S.R. (1989) *The seven habits of highly effective people: restoring the character ethic*, New York: Simon & Schuster.

Cowan, C. (1990) *Partnering: a strategy for excellence*, Phoenix, AZ: Department of Transport.

Cowan, C. (1992) *A strategy for partnering in the public sector*, ADFOT.

Dell'Isola, A. (1982) *Value engineering in the construction industry*, 3rd edn, New York: Van Nostrand Reinhold.

Friend, J.K. and Hickling, A. (1987) *Planning under pressure: the strategic choice approach*, Oxford: Pergamon.

Green, S.D. (1994) 'Beyond value engineering: SMART value management for building projects', *International Journal of Project Management* 12(1): 49–56.

Green, S.D. (1996) 'A metaphorical analysis of client organisations and the briefing process', *Construction Management and Economics* 14(2): 155–64.

Green, S.D. and Simister, S.J. (1999) 'Modelling client business processes as an aid to strategic briefing', *Construction Management and Economics* forthcoming.

Kelly, J. and Male, S. (1993) *Value management in design and construction*, London: E. & FN Spon.

Miles, L.D. (1972) *Techniques for value analysis and engineering*, 2nd edn, New York: McGraw-Hill.

Morgan, G. (1986) *Images of organisations*, Beverly Hills, CA: Sage.

Norton, B.R. and McElligott, W.C. (1995) *Value management in construction: a practical guide*, Basingstoke, Hants: Macmillan.

Parker, D.E. (1985) *Value engineering theory*, Washington, DC: Lawrence D. Miles Foundation.

Romm, J.J. and Browning, W.D. (1994) *Greening the building and the bottom line*, vol. 3, Old Snowmass, CO: Rocky Mountain Institute.

Romm, J.J. and Browning, W.D. (1995) 'Energy efficient design', *The Construction Specifier* (June): 44–51.

Rosenhead, J. (1989) 'Introduction: old and new paradigms of analysis', in J. Rosenhead (ed.) *Rational analysis for a problematic world*, Wiley, Chichester, Sussex: John Wiley, 1–20.

Turner, S.J. (1994) 'The philosophy of partnering in construction contracts honours thesis, Faculty of Engineering and Surveying, University of Southern Queensland, Toowoomba.

4 Organisational design

Mike Murray, Dave Langford, Cliff Hardcastle and John Tookey

Temporary project organisations

The work of Cherns and Bryant (1984) in identifying temporary multi-organisational dynamics has been widely recognised by several academic authors. Anumba *et al.* (1996), Luck and Newcombe (1996), Shirazi *et al.* (1996) and Winch (1989) have all discussed various aspects regarding project teams. Cherns and Bryant proposed 20 hypotheses which they acknowledged were subject to modification and qualification in the light of further evidence. Indeed, they recognised that very little research had been conducted in this area: 'with the appointment of consultants, contractors and subcontractors we are in the less well-charted waters of multi-organisational dynamics' (Cherns and Bryant 1984: 180).

More than a decade later Shirazi *et al.* (1996: 211) appear to re-emphasise this predicament, which suggests that the variables surrounding temporary project organisations remain uncharted waters, stating that 'an additional issue which is rarely addressed in the organisational literature is the temporary nature of the construction team and the fact that the team is generally made up from team members who work for different organisations'.

The assumption that all project teams are temporary may be incorrect. That is to say, anecdotal evidence suggests that the use of strategic partnerships between clients, contractors and subcontractors is growing at a steady rate and that this therefore eliminates the temporary nature of the project team. It is pertinent, however, to examine several of the hypotheses proposed by Cherns and Bryant. It is suggested that the following (re-numbered) points (as Shirazi *et al.* 1996 note) have implications for construction project organisational design with regard to the variables surrounding the temporary nature of the team.

[Point 1]: The management of a construction project from inception to completion is a function of a temporary multi-organisation (TMO) comprising relevant parts of those component organisations.

(Cherns and Bryant 1984: 181, point 2)

It is widely acknowledged that the move away from non-traditional procurement routes, especially towards management methods, has increased the number of organisations contributing to the construction process. The aspects of integrating and coordinating these organisations forms a central part of achieving successful project outcomes.

> [Point 2]: The actual performance of the [TMO] is determined more by the managerial capabilities of its component organisations and their co-ordination than by the form of contract [which is, in effect, the equivalent of the articles of association of the TMO].
>
> (ibid.: 182, point 5)

This point has relevance to Sir Michael Latham's 1994 report, *Constructing the team*, and its recommendation to use the Engineering and Construction Contract [E&CC, previously known as The New Engineering Contraact (NEC)]. The issues raised include the ability of work package contractors to cooperate with the lead contractor and amongst themselves for the benefit of the project objectives. It is suggested that the use of the E&CC will encourage trust and respect between the participants, something that Cherns and Bryant are unlikely to have observed with Joint Contracts Tribunal (JCT) type contracts in 1984. The other aspect concerns the management capabilities of the TMO's component organisations. This issue is currently being addressed, not by the contractors but by the industry's clients, who are unwilling, as they see it, to employ contractors with inefficient management practices:

> Among the issues identified by the Construction Round Table [CRT] and in need of urgent attention are: getting better site supervisors; stripping out some of the layers of management; and improving the quality of leadership and management within contracting firms.
>
> (*Contract Journal* 1998: 2)

This is as Cherns and Bryant predicted:

> [Point 3]: The earliest decisions taken by the client system have more influence over the way the [TMO] is formed and its subsequent performance than those taken later.
>
> (ibid.: 182, point 10)

This issue appears to have direct relevance to the views being expressed by both the CRT and the Construction Client Forum. That is, anecdotal evidence appears to support the hypothesis that clients are taking control of the organisational design of project teams and that this has clear implications for contractors and subcontractors attempting to meet client expectations.

[Point 4]: The performance of the [TMO] in construction management can be fully explained and understood only if the client organisation, which initiates and is part of the [TMO] is itself seen as a complex system with a past, present and future.

(ibid.: 182, point 15)

The need for contractors to understand fully the client's business operations and appreciate the reasons why they require a building is of paramount importance to many of the top construction clients (as argued in Chapter 3). An initiative by supermarket chain Tesco – Building the Future – highlights this: 'Tesco dumped its traditional approach to tendering every project. We needed to establish continuity. We wanted contractors to become an extension of Tesco and to understand us culturally' (*Building* 1997: 14).

Gardiner, in discussing the management of conflict in construction projects, also highlights this point. The following is part of a transcript from an interview with a client (on the City Technical College Project which was valued at £5.4 million):

We ran a day and a half orientation exercise before it even went to site to which the project manager, his leading foreman, the quantity surveyors and everyone else was invited and which everyone attended. The result was they knew what we were trying to build and what its purpose was. They were told about the National Curriculum, the way of teaching and about the concept of activity rooms and how the children would use them: so they had a very clear view, not just of the bricks and mortar, but of the purpose and why the timetable was so tight.

(*idem* 1993: 88)

The need for contractors to understand the client business is further exphasised by Barlow *et al.*:

Nevertheless, the client in the longer term partnering relationships [McDonalds, NatWest and Safeway] all stressed how important it was for their partners to learn about their needs and take on board their specific values – the embedded knowledge about the client.

(*idem* 1998: 91)

Jennings and Kenley examine the impact of human perception on the assessment, design and management of project organisations. They argue that the functionalist approach, which they suggest was prevalent in 1980s, has been shown to be incapable of comprehensively and accurately representing project organisations:

The physical nature of the building product is logically a powerful element in any person's perception. However, even in construction the management of the project organisation is primarily a social role; therefore the relative dominance of a technical perception for all aspects of project organisation is misleading.

(*idem* 1996: 239)

They argue for more explicit recognition of the non-technical determinants of project organisations and advocate the use of systematic thinking in designing and managing project organisations:

The systems interpretation is shown to be useful not only as a tool for facilitating understanding but also as a base for selecting appropriate management strategies to address given issues.

(ibid.: 240)

Jennings and Kenley consider the role of the temporary project organisation and use the concept of a building lifespan as a means to show how a project organisation evolves between the conception, construction and disposal of a building. They observe that the role of managing projects has evolved to be one of integrating specialist firms to form temporary multi-organisations (Davidson 1987; Mohsini and Davidson 1986). However, they acknowledge the difficulty in attempting to design the appropriate form and function of project organisations:

The construction industry's tendency to bring forward the previous phase's approach to the managing and integrating project organisations is commonly used, even though the phase may differ significantly in componentry and structure. This often causes the application of an inappropriate management strategy that may not only disregard significant technical but also social attributes of the project organisation's component firms.

(ibid.: 244)

With reference to the work of Green (1994) and Jackson (1991) they argue that the current mechanical/unitary or functionalist perception restrains the enhancement of non-mechanistic interpretations of project organisations by industry practitioners, causing, for example, professional, cultural and sentient conflict between firm and project objectives. Thus they conclude that the industry is preoccupied with the classical school of management which perceives systems or organisations as machines and expects participants to act in a unitary manner in their organisations. The systematic qualities of project organisations (human and cultural issues) can only continue to become more important because of the inherently social role of integrating specialist organisations within project organisations. Moreover, without recognition of these social needs the assurance of project organisation's performance will become increasingly difficult.

Team building, empowerment and integration of the project participants

Team building

> Teamwork involves the effective cooperation of a group of people in activities that are directed towards a common goal. The whole point about teamwork is that the performance of the group as a whole is better than would be the sum of the performance of the individuals composing the group.
>
> (Gabriel, 1991: 196).

Gabriel identifies the benefits derived from synergy and suggests that there is a limited lifespan for which the synergy of a team can be maintained. (Nature abhors a perpetual state, and the maintenance of team effort is no exception). It is suggested that three years is the maximum time that a team approach can be maintained; something which, if valid, has implications concerning the role of strategic partnering arrangements in construction projects.

Rolstad (1991) examined the benefits derived from team-building activities during project start-up. He concluded that the challenge is to create the right organisation and organisational spirit to face the considerable tasks and challenges in the project and not least to bring all those individuals with varying backgrounds and experience into one team with the common goal of providing the agreed product and service on schedule and within budget.

Gray and Suchocku (1996) suggest that many of the existing techniques for team building are inappropriate for the construction industry and that empirical evidence shows that workshops based on project start-up are effective but that they are rarely used. Their case study showed that people were poorly introduced to their specific roles on the project and would clearly like to be properly involved through induction. Sommerville and Dalziel (1997) draw attention to the use of psychometric testing of team members using the Belbin Team-role Self-perception Inventory approach (Belbin 1981). This technique may be appropriate for the processes involved in team development and composition.

Gardiner (1993) also refers to the use of psychometric type instruments which he suggests can aid clients and project managers when selecting team members. Gardiner conducted interviews with several construction clients and the data collected clearly showed an increasing concern by clients and their advisers to assemble a team which will function well together. Indeed, the following extracts from the transcript of the interviews with two clients clearly highlight this concern:

> On our design and build contracts we even state that we have the right to say who the agent is, or whom we do not want. Once he is on the job

the contractor cannot take him off, or move him about without prior comment.

(ibid.: 107)

We also assessed the contractors by the team they put forward. We insisted on seeing the team, the actual people running it. The chemistry needs to work between the people involved. I wouldn't say that we are experts on assessing individuals but you do form your own prejudices and opinions.

(ibid.: 107)

Tampoe (1989) believes that project success can be obtained by the concept of the empowered team member whereby the team takes responsibility and is accountable for meeting the project goals. This view tends to emphasise the need for less control by the project manager but requires three pre-requisites for success. Figure 4.1 highlights the need for the organisational climate together with employee competence and commitment to be correct to achieve optimum effectiveness from an empowered team.

Empowerment

The negative consequences that power and politics can have on projects is examined by Newcombe in relation to a compromise of two procurement paths: traditional and construction management. He suggests that:

Construction Management is based on the modern management principle of empowerment or power equalisation and reflects the trend towards a more pluralistic society. Whilst ultimate authority rests with the client, as indeed it should, the encouragement of participation by all parties in decision making coupled with the democratic distribution of power provides a radical new basis for conducting construction projects. Current management thinking [Peters 1987, 1992; Kanter 1983, 1989; Morgan 1993] suggests that this is the only way to run successful organisations now and in the future.

(*idem* 1994: 527)

Newcombe (1997) concludes that a strong positive culture is more likely to be present in procurement paths that encourage empowerment and participation – construction management – than in those that engender fragmentation and friction, for example in the traditional system (power being based on a hierarchy of command with an authoritarian approach to managing).

The concept surrounding empowerment, as shown by Tampoe (1989) and Newcombe (1997), would appear to have been present in the Broadgate

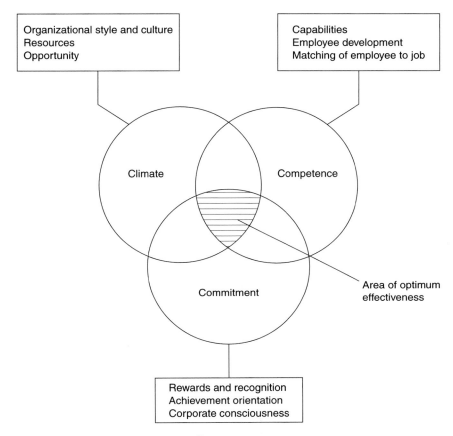

Figure 4.1 Empowering project staff
Source: Tampoe 1989
Reproduced by kind permission of *International Journal of Project Management*, an imprint of Elsevier Science Limited.

Office development project in London (undertaken between 1985–91). Table 4.1 demonstrates the culture which was fostered on the project and is described as one of the industry's most striking success stories (*Building* 1998).

Integration

Ahmad and Sem (1997) developed a contingency model which involved an iterative approach of designing construction project teams in order to minimise the negative effects on total quality management (TQM). They link the aspects of team design, project characteristics and project success factors (TQM) and provide a useful example of viewing the organisational

Table 4.1 Empowerment at Broadgate: the Broadgate culture

There was a project-wide culture of openness, where problems could be resolved quickly, with:	Broadgate was a forerunner of supply-chain management, where specialists and trade contractors are invited to share in keeping down costs:
monthly meetings for the directors of all companies involved	subcontractors were flown to the USA to learn time-saving and cost-saving techniques established there
quarterly meetings with union officials	
daily 7:30 a.m. 'toolbox' meetings between Bovis or Schal section managers and tradespeople (an idea taken from Japan)	steel erectors were equipped with tools previously used only in the USA
	hoist manufacturers were invited to design larger hoists that could carry bulky pallets of plasterboard without damaging their cargo
induction meetings to explain the scope and aims of the project to trade contractors and off-site manufacturers	
regular meetings with all tradespeople in the site canteen	prefabrication and pre-assembly of components such as toilet pods and air-handling units were maximised
a reward culture fostered through monthly prize-givings.	plasterboard manufacturers were encouraged to take the simple measure of wrapping different types of board in different coloured paper to avoid the wrong quality of board being installed.

Source: Building 1998

design process. Table 4.2 shows project team contingency factors, and Figure 4.2 indicates how these are dependent on the project characteristics which have an impact on TQM. The role of the parent organisation client is also examined, and Ahmad and Sem suggest that the factors present in the project team are likely to mirror those of the parent organisation. However, they conclude that this may not always be the case, and even a mechanistic parent organisation with an autocratic leadership style can form an organic project team with democratic leadership.

Table 4.2 Project team contingency factors with extreme possibilities

Factors	*Extreme possibilities* ←——— *continuum* ———→	
Structure	Mechanistic	Organic
Control system	Formal or market	Clan (trust)
Leadership style	Autocratic	Democratic
Values	Stability, risk averse	Innovative, risk taking

Source: Ahmad and Sem 1997
Reproduced by kind permission of E & FN Spon

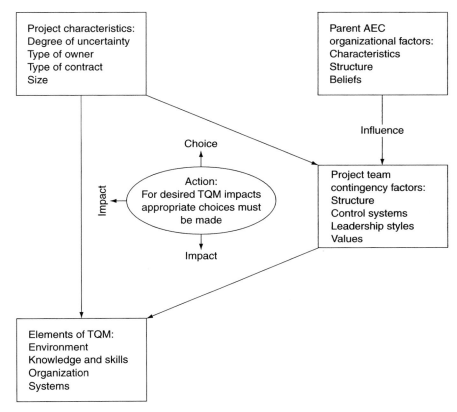

Figure 4.2 Factor element impact model for forming total quality management (TQM) oriented construction project teams
Note: AEC = architectural, engineering and construction enterprise
Source: Ahmad and Sem 1997
Reproduced by kind permission of E & FN Spon

Partnering arrangements and project stakeholders

The concepts of partnering and project stakeholders are examined here as a means of looking deeper into the topic of project teams. Research carried out by the Chartered Institute of Building (*Construction Manager* 1998) suggests that the construction industry is 'holding its breath' on benefits to be derived from this cultural shift of avidly adopting partnering, in all of its guises. Table 4.3 shows that several of the top 20 UK construction companies had partnering arrangements in 1998 and intended to increase the number of such arrangements through 1999. There is no indication whether the tactics behind such enthusiasm are rooted in commercial gain or a deep desire to operate in a non-adversarial environment.

The degree to which partnering may alter the decision-making process

Table 4.3 Contractors' partnering workloads

Company (%)	Contracts involving partnering 1998 (%)	Contracts to involve partnering in 1999
Alfred McAlpine & Sons Ltd	60	fna
Amec PLC	50	fna
Amey Construction	60	100
Ballast Wiltshier Construction	20	25
Birse Constructions Ltd	40	fna
Bovis	50	fna
Clugston Construction	10	15
HBG Construction Ltd '	50	65–70
John Laing PLC	20	fna
John Mowlem & Company	26	45
Kier Group Ltd	25	fna
Mace Construction	20	35
Miller Group Ltd	70	fna
Sir Robert McAlpine & Sons Ltd	40	50
Tarmac PLC	20	fna
	33	50
Tilbury Douglas Construction	20	30

Source: *Construction Manager* 1998
Note: fna = figures not available, but increasing

regarding organisational design is observed by Barlow (1997), who states that 'In some instances there had been an explicit attempt to reform a company's organisation structure to produce cross-functional teams (eg Simons Construction, contractors to NatWest Bank and Safeway)'. Barlow *et al.* conclude by discussing the benefits of organisational learning that can be derived from partnering arrangements. They specifically note that

> This study shows that currently the main driving force in the typical construction industry partnering relationship is efficiency improvement, but clients and suppliers are beginning to recognise the role partnering can play in promoting innovation and learning at an individual team and organisational level.
>
> (*idem* 1998: 12)

One aspect of partnering, namely trust, which is necessary to promote a culture of learning throughout the partnership, is examined by Crowley and Karim (1995). Their model graphically illustrates partnering concepts, which include different patterns of communication flow, modified company boundaries and joint alliances within the project organisation, as shown in Figure 4.3. The greater the number of organisations involved within the partnership the more complex the organisation design and communication patterns become. However, the benefits are that the shared culture creates a

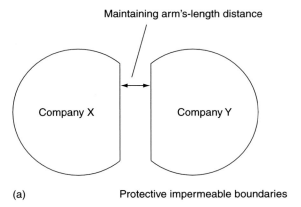

Maintaining arm's-length distance

Company X

Company Y

(a)   Protective impermeable boundaries

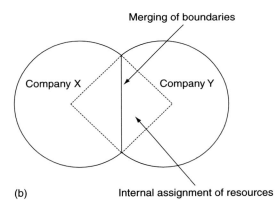

Merging of boundaries

Company X

Company Y

(b)   Internal assignment of resources

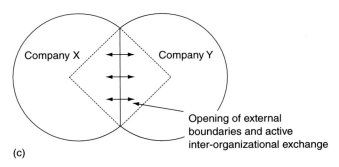

Company X

Company Y

Opening of external
boundaries and active
inter-organizational exchange

(c)

Figure 4.3 Growth of a partnering relationship: (a) traditional relationship; (b) formative partnering stage; (c) partnering relationship with permeable boundaries

Source: Crowley and Karim 1995

Reproduced by kind permission of the American Society of Civil Engineers synergistic atmosphere to blend the objectives of the participants into a greater whole and to incorporate their ideas and resources.

Larson and Drexter (1997: 46) ask, 'What are the major barriers to successful partnering in construction projects?' They note that 'Partnering represents a "paradigm shift" in how one approaches construction projects'. Their study revealed that a general level of mistrust existed between owners and contractors which was engineered by years of viewing and treating each other as potential adversaries. Table 4.4 shows themes and subthemes that were identified by project managers, principal engineeers, consultants and specialists as the major barriers to successful project partnering. It is worth examining several of these themes as it must be expected that the perceptions identified are likely to exist on any project, whether partnered or not. They are therefore contingencies which must be taken into account when considering the organisational design of a project. Some 31 per cent of the respondents suggested that a failure to build a true relationship of trust was significant and would prevent successful partnering taking place. Another significant problem identified involves the difficulty encountered in synchronising goals in organisations where numerous subcontractors are used. Clearly, the issue of integration and coordination are of paramount importance both for partnered and non-partnered projects. Other barriers that were identified included the complications which arise from the differing values and culture of the people involved in the process and the inability of 'Jurassic Park' management to allow change to take place.

The decision to participate in a project which involves partnering (between client and contractor, contractor and subcontractor, etc.) allows for the examination of the design of project organisational structures to be viewed under different assumptions from those of non-partnered projects. The degree to which authority, power and decision-making differ in these types of projects is therefore an important factor determining success.

Redesign of project organisations

Introduction

Two recent research projects are examined here in the belief that the proposals presented, by Higgin and Jessop (1965) and by Cox and Townsend (1997), will require the parties involved in construction projects to participate in a 'paradigm shift'. That is, the proposals move the current discussion requiring the design of organisational structures for construction projects into a new 'arena' where the design variables may require adjustment.

The Tavistock Institute is currently engaged in coordinating the redesign of interorganisational arrangements (Figure 4.4) in two demonstration building projects. The redesign involves the concept of 'work clusters'. In each work cluster the designers, subcontractors and key suppliers involved in a reasonably self-contained element of the building undertake a form of

Table 4.4 Barriers to successful partnering

Type of barrier	Frequency stated		Characteristic comment or difficulty
	number	per cent	
Trust, attitude and interpersonal barriers:			
trust	58	31	Failure to build a true relationship of trust
past adversarial relationships	19	10	Many people with an instinctive suspicion of the other party owing to past experiences
fear	6	3	Fear of the unknown and change
attitude	19	10	Differing values and culture of the people
owner–slave mentality	9	5	The owner–slave concept is alive and doing well
ego	15	8	Some people have an ego problem . . . they cannot admit there is another way
old-fashioned management	20	11	Old habits die hard
greed	16	9	Money: there is never enough. When there is 'extra', greed seems to get involved
risk-sharing	11	6	A lack of understanding risk and how it is redistributed in a partnering environment
profit motive	10	5	Not understanding fair profit
communications	14	7	I would be concerned that information divulged would be used [against us] later
validity of partnering process	7	5	Non-acceptance [of partnering] as a long-term way of doing business
Project structure barriers:			
contracts	18	10	[We are] too reliant on legal protection and advantages or loopholes in documents
government and legal issues	24	13	Public projects require complete documents and a fixed price, precluding full partnering
low-bid structure	13	7	Low-bid method of awarding projects . . . leads to built-in conflict
international bureaucracy	3	2	Multi-layer administrative management with little involvement and lukewarm support
finding a partner	3	2	Finding contractors who are able to partner, so that the owner benefits
not appropriate	9	5	Formal process is suitable only for large projects

Table 4.4 Continued

Partnering process barriers:			
lack of common goals	32	17	Synchronising goals in a [big] organisation or when numerous subcontractors are used
time constraints	9	5	Time and effort required to align both organisations with every undertaking
poor planning	18	10	Failure to use plan. Partnering requires management by planning
expensive	15	8	It is extremely costly and inefficient
Knowledge and skill barriers:			
not understanding partnering	23	12	Unfamiliarity or misunderstanding of partnering concepts by upper management
no experience with partnering	5	3	Experience in this type of approach to contracting
knowledge of other's business	13	7	Lack of understanding of the partner's company's culture
micro-management	16	9	The other party does not stand behind their engineers' decisions
Commitment:			
commitment	19	10	Failure to 'walk the talk'
top-management commitment	19	10	Senior management does not show support

Source: Larson and Drexter 1997
This figure is reprinted from *Project Management Journal* with permission of the Project Management Institute, 4 Campus Boulevard, Newton Square, PA 19073–3299, a worldwide organisation of advancing the state-of-the-art in project management.

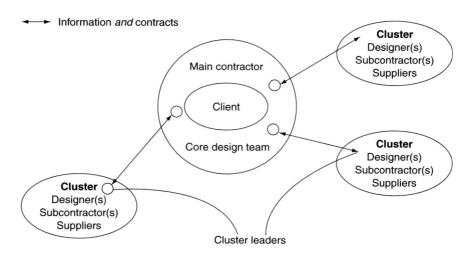

Figure 4.4 New interorganisational arrangements
Source: Tavistock Institute 1996/97

simultaneous engineering. Typical clusters include groundwork, frame and envelope, swimming pools, mechanical and electrical services and internal finishes. Within each work cluster the participants use techniques including value management and risk management and contribute knowledge to anticipate the areas of interdependence which are likely to arise during detailed design and construction.

The clusters are responsible not only for designing but also for delivering construction of their element of the building. Interdependencies that span the spheres of two or more clusters are resolved by taking them to an overall project team, where each cluster is represented by a cluster leader. The concept of work clusters is an attempt to resolve the problems identified in earlier Tavistock Institute research (Higgin and Jessop 1965) which concluded that the design and construction of buildings is characterised by high levels of interdependence and uncertainty.

It is suggested that the cluster model will improve the whole design and construction process by clearly identifying areas of interdependence and thereby reducing uncertainty. To some extent, this type of project framework already exists within the industry, especially with projects procured under management methods. Subcontractors are increasingly involved in the design process with regards to their input at early stages of the project concerning 'buildability' issues.

The Tavistock Institute (1996/97: 48) recognise that the work-cluster method requires the adoption of long-term (strategic) partnering arrangements to be developed to ensure the viability of its proposal: 'This means that Main Contractors must be willing to invest in developing some form of partnership arrangements with the key subcontractors and suppliers beyond the span of an individual project'.

The work-cluster arrangement also provides an interesting vehicle for examining the structure, decision-making process and communication patterns of a construction project. It is possible, however, that by examining current projects the work-cluster arrangement may indeed already exist, albeit under a different guise.

Research conducted by Cox and Townsend (1997) for British Airport Authority (BAA) differentiates between clients that have a regular requirement for construction work of a similar value and content (process spend) and clients that make infrequent purchases (commodity spend). They propose that 'collaborative' and 'teamwork' approaches as suggested by Latham (1994), are possible only when there is a long-term relationship based on regular spending (that is, on process spend).

Cox and Townsend's proposal for a (process-spend) industry structure is shown in Figure 4.5. They note that it resembles an induced quasi-vertically integrated supply chain somewhat reminiscent of the Japanese Keiretso but not as tightly interlocked. The overall effect was to create a pool of contracting firms with a relatively stable demand for their services and consequently, a reasonably secure income. This would in turn lead to contractors

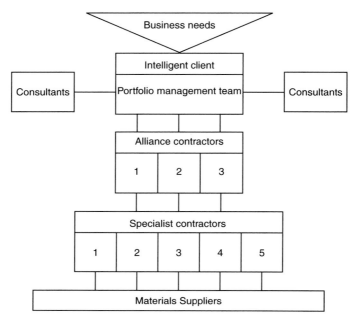

Figure 4.5 Anticipated industry structure (process spend)
Source: Cox and Townsend 1997
Reproduced by kind permission of Blackwell Science Ltd

establishing a 'fit for purpose' long term relationship with their own sub-contractors and suppliers. The benefits from adopting this model include a reduction in fragmentation in the construction process and an erosion of functional and organisational barriers as a result of striving to achieve common goals (satisfying the client needs).

Environmental influences and organisational design of projects

Sidwell (1990) explores the nature of project management in complex dynamic environments. He examines the work of Child (1972) and Mintzberg (1979) and notes that an organisation's environment can range from stable to dynamic and that the higher the variability and uncertainty of that environment then the more the organisation needs to be organic and adaptive. Indeed, Sidwell recognises that construction project environments are usually dynamic, complex, diverse and hostile and that this may occur across all components of the environment and may do so simultaneously.

Sidwell discusses the contingency theory of organisational design and comments that the effectiveness of the project organisation depends on a range of diverse factors such as the technology of the project, its size, the project environment, roles and relationships of team members and the degree of management control. Moreoever, the construction project team is a living organism; at each phase in the project life cycle it transforms in structure and style. He suggests that individual firms have to balance the goals of the project with their organisational goals and must cope with the vertical and project matrix structures and styles of management.

An examination of the work carried out by Cleland and King (1975) in relation to matrix structures is undertaken, and Sidwell also reiterates the view of (Bennett 1985) that 'The relationship between functional, matrix and project organisations can be represented as a continuum with functional and project organisations at opposite ends with matrix lying approximately midway between the two ends'. (Sidwell 1990: 163)

Sidwell moves on to compare and contrast modestly-sized projects with large complex projects and notes that the former are generally categorised by reduced communications and coordination problems. He proposes a pictorial model which illustrates the style changes that can take place during a project's life cycle (see Figure 4.6). He observes that:

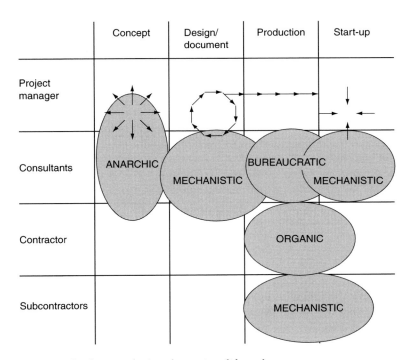

Figure 4.6 Style changes during the project life cycle
Source: Sidwell 1990
Reproduced by kind permission of E & FN Spon

What we see in large project teams is a nesting of organisational approaches. The larger structure develops bureaucracy in order to cope with the volume of business and yet it must retain its organic, responsive nature at the workface. Because projects have a distinct life-cycle there is also a horizontal time related dimension to their organisational form, they change over time from organic to mechanistic to bureaucratic.

(ibid.: 165)

Sidwell uses case studies of major projects to demonstrate his theory and concludes that in the projects examined the project and construction management organisational forms had the ability to evolve and embrace the multifaceted management structures, organisations and styles necessary to cope with the dynamics of these major projects. Indeed, it is the project organisation's ability to metamorphose within the project environment in consideration of the chosen procurement route which is likely to lead to a successful project outcome.

Shirazi (1990) develops models based on established approaches to organisational analysis and decision-making processes as a means of determining the impact of environment and technology on the formal structures of site organisations. The five models all have differing assumptions and biases and were tested on 18 construction and civil engineering projects as a means of identifying the one that dominated the task of structuring. The models are based on the work of Tatum (1984) who in turn notes that his own investigation used five models of decision-making distilled from existing theory. Shirazi's five models were the adaptation model, the behavioural choice model, the organised anarchy model, the political model and the contingency model and are used to help to group certain recurring patterns and to allow for the formulation of decision processes at the earliest phase of organisational structuring.

Shirazi refers to behavioural decision theory and examines March and Simon's (1958) definition of organisation: 'a structure of decision making acting at times as individuals and at other times as groups'. He observes that the behaviour view argues that decisions are made under conditions of bounded rationality which implies that individuals make decisions under a number of external and psychological constraints (Cyert and March, 1963). He uses the work of Mintzberg (1979) in categorising four broad groups of organisational structures – design of position, design of superstructure, design of lateral linkage and design of decision-making systems – which are formal and semi-formal means of grouping and coordinating organisation activities to establish distinct patterns of behaviour and therefore affect how an organisation functions. He concludes:

1   A shift in environmental or technological conditions leads to a disequilibrium which produces a decline in effectiveness and creates a

pressure for change, resulting in structural adaptation to a new structure and thereby restoring effectiveness.

2   A move towards a rational process in decision making and a consistency among contingency factors and the design parameters is often dependent on the available data regarding the situation and time.

3   The dominant process in structuring appears to be frequent adaptations of past structures, particularly when quick responses are sought and explained in terms of interactions of deterministic and voluntaristic processes to retain a preferred organisational configuration.

Shirazi *et al.* (1996) note that the temporary nature of the project team (composed of independent organisations) leads to a less bureaucratic and more decentralised structure than would be predicted using conventional organisation theory. However, they observe that temporary multiorganisations appear to follow the prescriptions of systems theory in clearly differentiating between the managing and operating system (see also Walker 1997).

Other authors who have examined organisation structures on construction projects include Hughes (1989), who developed his model based on the work of Walker (1980). Indeed, Hughes observes that the strength of Walker's approach lies in its diverse origins (that is, it is based on the works of Burns and Stalker 1966; Lawrence and Lorsch 1967; Thompson 1967; Cleland and King 1975 and others). However, Hughes is critical of Walker's model in that although he observes that his linear responsibility analysis (LRA) is the most comprehensive graphical portrayal of temporary organisational structures, it is the very comprehensiveness which renders it too unwieldly to use for large projects. Moreover, he notes that Walker's definition of the environment is too loose and gives little scope for making systematic connections between the environment and the organisational form. Hughes suggests that any observable environmental phenomena surrounding a project can be classified into one or more generic groups of environmental forces (that is, political, legal, institutional, cultural, social, technological, economic, financial, physical, aesthetic policy). These generic groups (drawn from the work of Hodge and Johnson 1970; Walker 1980; Farzad 1984; Kast and Rosenzweig 1985) are used by Hughes (1989) in a brief study of the UK and Jamaican industries.

Hughes uses case studies (public sector projects) to test his '3R' model (roles, responsibilities, relationships) which he suggests meet the need for a technique to describe systematically and to evaluate quantitatively the organisation of building projects. In combining the lessons learned from Walker's study of private sector projects Hughes suggests the following steps be taken when setting up a construction project organisation:

1   The organisation structure should be designed at the outset of the project, using '3R' charts.
2   The objectives of the project should be defined in terms of the effect intended on the environment.
3   Policy decision points should then be identified.
4   Within each policy system, strategic decisions should be identified.
5   Within each strategy subsystem, tactical decisions and operations should be identified.
6   Responsibilities for the operations should be defined and opportunities for input from the end users of the building identified.
7   The request level of control should then be superimposed on the responsibilities, relating to the achievement of the objectives defined in Step 2. The management structure is thus identified.
8   Co-ordination, integration, feedback and client involvement patterns should be identified, and mapped on to the set of '3R' charts which describe the organisational structure of the project.
9   From the columns of the '3R' charts, job descriptions for each of the members of the project team can be identified and used for the selection and the appointment of the consultants etc.

(ibid.)

Hughes accepts that his study has limitations and that it focuses exclusively on the structural aspects of organisations and therefore little was learned about the effects of behaviour on organisational effectiveness. He does, however, refer the reader to the work of Williamson (1981) who examined the 'human relations' problems regarding the differentiation and integration issues of constituent organisations within a project.

The use of a systems approach is proposed by Moore and Moore (1997) for examining the manner in which subsystems within a project determine the extent of openness required by an organisational structure. They suggest that the common perception of the construction industry as being typically low-profit with a high failure rate stems largely from the failure of the industry to provide relevant organisational structures for its projects. They present a simplified representation of the relationship between a project environment and organisational structure (Figure 4.7) which combines the work of Hughes (1989), Mintzberg (1979) and Duncan (1971).

Moore and Moore observe, however, that a more structured approach is required to carry out an analysis of their hypothetical project. They extend the environmental factors used by Hughes (1989), which involved quantifying the project environment under two dimensions (mitigable–unmitigable; definable–undefinable).

The new model developed takes account of Osborne and Hunt's (Woodward, 1961) views regarding the differing levels of intensity (degree of influence) acting on the project's environmental subsystems. In addition, the work of Walker and Kalinowski (1994) in developing Osborne and

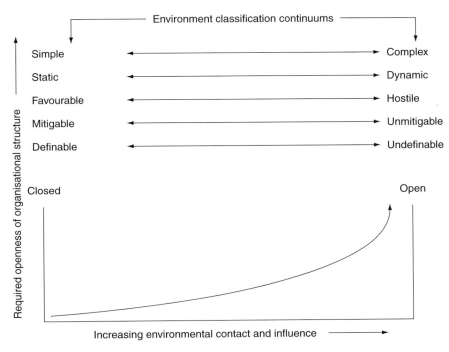

Figure 4.7 Simplified representation of the relationship between project environment and organisational structure
Source: Moore and Moore 1997
Reproduced by kind permission of the Associate of Researchers in Construction Management

Hunt's concept of micro and macro environments is also developed further in their model. This involves creating a new third level, the meso level, to take account of the hypothetical project used in the analysis. This is deemed necessary where management contracting procurement methods are used. Moore and Moore argue that the works contractors are not entirely controllable by the project, but are contained within it. They suggest that this argument can be viewed as an extension of Shirazi *et al.*'s (1996) conclusion that project organisation structures act as a buffer between the internal and external environments of the project. Figure 4.8 shows the relationship between each of the levels and the continua over a single environmental subsystem, that is, cultural, economic, political, etc. Moore and Moore conclude that the system representing the hypothetical project is 'homeostatic' (one capable of a certain level of internal adjustment in order to subjugate the effects of the external environment) whilst tending towards being open. Their model, it is suggested, will also give realistic consideration to the key subsystems within a project prior to the designing of an appropriate organisational structure.

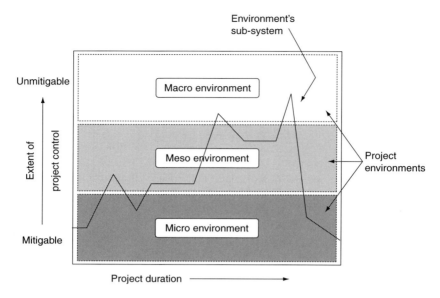

Figure 4.8 Suggested project environments and varying levels of control over a
single environmental subsystem
Source: Moore and Moore 1997
Reproduced by kind permission of the Associate of Researchers in Construction
Management

Winch (1989) reviews three perspectives for analysing construction
management – socio-technical systems; organisation and environment; and
project management. Winch concludes that in spite of their considerable
usefulness, they contain no framework for analysing the inevitable differences in interest between the different firms who are members of the project
coalition.

Winch introduces the reader to the concepts of 'transaction costs' and
'hierarchy' which are derived from the work of Williamson (1985) and
argues that to some extent these overcome the deficiences of the three
perspectives noted above. Winch refers to Reve and Levitt (1984) who
define transaction costs as costs resulting from the additional layers of
managerial hierarchy and specialised staff in construction projects who
maintain elaborate surveillance and control systems. Winch notes that the
rational response to project uncertainty, project complexity and the post-
contract bilateral monopoly situation identified by Reve and Levitt would,
according to a 'transaction cost' analysis, be 'hierarchy'. Winch suggests
that hierarchy would:

> Economise on bounded rationality due to uncertainty and complexity.
> In particular, the designer/specialist subcontractor transaction inter-
> faces could be beneficially governed by hierarchy rather than market. In

essence, the advantages of hierarchy for the construction process are the possibility of transfer of organisation and expertise from project to project and the facilitation of feedback loops from the construction to the design process.

(idem 1989: 341)

Winch asks why the process of hierarchy and the ability to transfer an established project organisation from one project to another has not been widely adopted in the British construction industry. He observes that the trend during the 1980s was indeed towards greater market governance with the emergence of management contractors and the retreat of main contractors from actually doing any site work. He suggests that a strategy based on market governance

Tends to undermine any attempts to reduce project uncertainty and to improve project productivity. Moreover strategies which improve the performance of construction firms in terms of profit maximisation are incompatible with the effective management of the project and strategies to improve the provision of the built environment.

(ibid.: 342)

Winch concedes that, in the light of his research, the transaction cost approach of Williamson designed to handle relations within and between firms did not explain why market governance still ruled the construction industry. The answer to this question he observes is found in the contradictions between project and contracting uncertainties and the institutionalised and deep-rooted nature of the professions in the industry.

Communications and decision-making during the construction phase

Communications between parties at the design stage of a project is a key area of focus. Bowen and Edwards (1996), for example, examine the communication process in relation to cost planning. Gameson (1996) conducts research into the interaction process between clients and construction professionals during the early stages of project development. Hardcastle (1992) develops an information model of the construction cost estimating process.

Thomas *et al.* (1998) examine the critical communications variables present within engineering and construction projects. They document the investigative efforts of the Construction Industry Institute Project Team Communications research team whose objective was to develop a diagnostic tool for improving project team communications. Their study established a positive and quantifiable link between communications effectiveness and project success. Table 4.5 shows the critical communications variables subdivided into six categories. Thomas *et al.* suggest that these variables can be used to measure the effectiveness of communications on other projects and can provide a basis for improving team communications. They observed that barriers and filters restricting communication flow may be the most difficult category to improve as they may be the result of interpersonal problems which are beyond the ability of the project manager to solve.

The difficulties in designing communication frameworks for construction projects which involve 'fast-track' principles (construction activities start before design is completed) and involve the integration and co-ordination of several work package contractors is analysed by Pietroforte (1997). He notes that the information process of a building project is not only a matter of describing requirements and translating them but also of communicating these requirements to different entities, making sure that their meaning is understood and that these requirements are fulfilled accordingly. He draws attention to the temporary nature of the project team and points out that the individuals involved tend to value structure and interpret the same information content in different ways.

Pietroforte introduces the concept of transactional governance which is used to explain the cost associated with coordination and communication effort whether within the same organisation or between different organisations. He explains the work of Coase (1937), Williamson (1979, 1985) and Boisot (1983) on hierarchies, markets and federations. Figure 4.9, adapted from Boisot (1988), is used to summarise the main informational and social features of markets, hierarchies and federations and shows their possible position within the codification and diffusion space of information.

Table 4.5 Critical communications variables

Accuracy:
The accuracy of information received, as indicated by the frequency of conflicting instructions, poor communications and lack of coordination

Procedure:
The existence, use and effectiveness of formally defined procedures outlining scope, methods, etc.

Barriers:
The presence of barriers (interpersonal, accessibility, logistic/or other) interfering with communications between supervisors or other groups

Understanding:
An understanding of information expectations with supervisors and other groups

Timeliness:
The timeliness of information received, including design and schedule changes

Completeness:
The amount of relevant information received

Source: Thomas *et al.* 1998
Reproduced by kind permission of the American Society of Civil Engineers

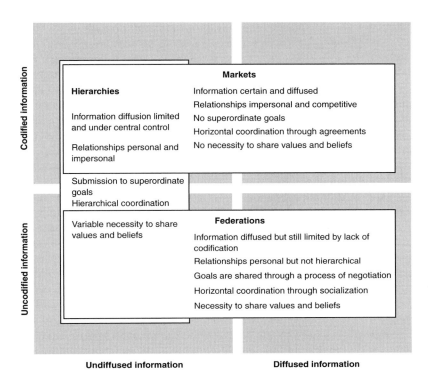

Figure 4.9 Hierarchies, markets and federations
Source: Pietroforte 1997
Reproduced by kind permission of E & FN Spon

Pietroforte concludes that building projects are successfully completed through the development of federative mechanisms, such as cooperation, informal rules and rules that complement and circumvent the hierarchical and formal provision of standard contracts. Moreover, the interactive nature of the design and engineering process, the growing number of participants and the interdependence of building systems demand increased informal communication and mutual adjustment as the coordination mechanisms.

The use of information technology (IT) as a means of communication is reviewed by Pietroforte. He observes that IT applications have been geared more towards the processing of accurate or quantitative data for routine tasks, such as job costing, scheduling and estimating, than the improved communication of qualitative and incomplete information across the organisations involved in non-routine design and engineering tasks. He recommends that, ideally, electronic links with capabilities such as interactivity, simultaneous two-way information exchange and flexibility of communication format should network all project participants despite their geographic dispersion and time limitations. Walker and Rowlinson discuss the impact of IT on the procurement system in Chapter 8.

Mead (1997) suggests that the development of project-specific intranets will enhance the information flow between project participants. He concludes that in an era of increasing complexity, these intranets may improve communication and improve teamwork on rapidly developing construction projects. Mead refers to a US construction management consultant that has developed intranets to form virtual environments for construction teams (see Chapter 8 for a further discussion of IT and procurement systems).

Mead explains the concept behind the project intranet as being an intranet e-mail distribution system that can be used to transfer communications and files between team players. The project intranet moves beyond previous models and includes a library of integrated management information that, it is suggested, can enhance the performance of the project team. A third model proposed by Mead (the comprehensive intranet model) involves a complete re-engineering of the construction information process. Figure 4.10 shows this model, which consists of four main libraries: project, management, design and financial information. These libraries become information 'hubs' that are maintained and controlled by specific teams within the project organisation.

Research into the communication process between contractors and subcontractors would appear to be limited. However, research undertaken by Matthews *et al.* (1996) suggests that most subcontractors believe that relationships can be improved between these parties if the communication process is improved.

The problems associated with the lack of communication and coordination in connection with reinforced concrete structures are examined by Shammas-Toma *et al.* (1998). The paper presented has two purposes: to examine the problems identified in earlier research regarding reinforced concrete structures and to consider how the findings from this research is explained. The earlier research found that:

> It was recognised by many participants that communication and co-ordination have to be improved between the parties to the project, so as to enable them to solve the problems arising from what they saw as the evident uncertainties of work on site. In addition it was also recognised that, in a temporary coalition of differentiated economic interests, there was no unified control and the attempts to achieve co-ordination through contractual formularisation and regulation was also failing. Findings suggest that contractual procedures constricted communication between the Resident Engineer and the subcontractor.
>
> (ibid.: 183)

In many cases informal procedures were used to overcome the contractual formalities in that the resident engineer would communicate with

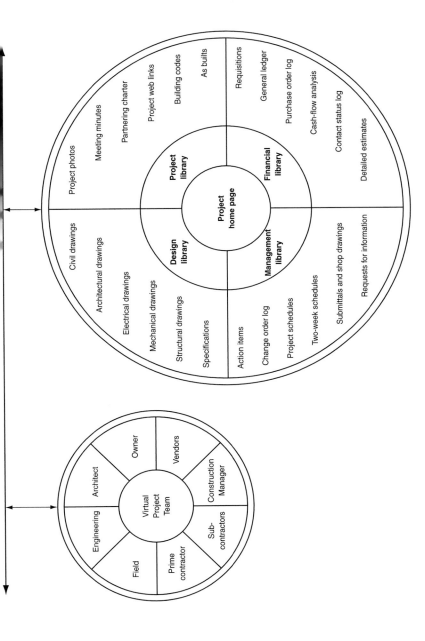

Figure 4.10 The comprehensive intranet model
Source: Mead 1997

This figure is reprinted from *Project Management Journal* with permission of the Project Management Institute, 4 Campus Boulevard, Newton Square, PA 19073–3299, a worldwide organisation of advancing the state-of-the-art in project management operatives to overcome a problem on-site as it arose without resorting to formal methods. However, they also observed that this may be interpreted as a breach of contract. This point is, however, re-emphasised in that suggestions from participants indicate that:

> Instructions to operatives should flow from once source, namely the contractor. The receipt of instruction from two sources may result in confusion. However, the gain in efficiency from enabling the Resident Engineer who is more fully aware of the design requirements would seem to be obvious to them. It may be concluded that contractual procedures are needed to provide communication channels between the Resident Engineer and operative.

(ibid.: 183)

Organisational design and project success factors

The role of project management in contributing to project success is examined by Morris (1989) who provides a diagram (Figure 4.11) explaining the factors necessary for the success of major projects. Although the study conducted by Morris focuses on 'mega projects' (The Channel Tunnel and the NASA Space Station) his conclusion has relevance for construction projects of any size. These are: the need to 'manage' external forces acting on a pro-

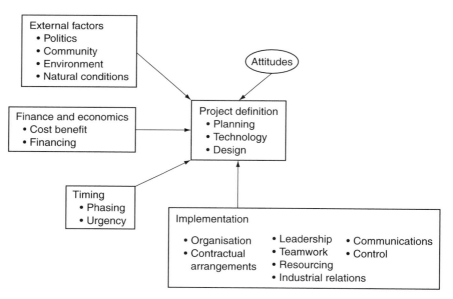

Figure 4.11 Factors affecting project success
Source: Morris 1989
Reproduced by kind permission of *International Journal of Project Management*, an imprint of Elsevier Science Ltd

ject (see Hughes 1989 on the environment of construction projects, for example) and the absolute importance of positive definition (workable in technical, financial and schedule senses) within an appropriate organisation which has adequate support.

The relevance of matrix forms of project organisation to project success are discussed by Larson and Gobeli (1987). Before examining their work it is pertinent to refer to the research undertaken by Bresnen (1990) who comments on the applicability of constructs and models derived from the investigation of complex project systems. He observes that the majority of research conducted into matrix management does not take account of factors particular to construction projects – the greater relevance of inter-organisational relationships. He suggests, however, that 'The small scale, less technologically advanced setting of a construction project may provide suitable opportunities for the fuller exploration of matrix management forms'. (ibid.: 13). He concludes by making a general point:

> By effectively treating the construction industry as a special case, important insights into contemporary organisations may well be lost. Whether the case of construction does indeed prove to be an exception to the rules after further consideration, or whether it in fact contributes towards further understanding of those rules still remains to be seen.
>
> (ibid.: 15)

In Larson and Gobeli's study success was gauged according to what they term 'traditional criteria' – cost, schedule, technical performance as well as overall results. They observe that as the interrelationship between these criteria varies the assessment of performance is somewhat subjective, that is 'Often projects are termed a technical success despite being between schedule and over budget. Conversely, projects may be ahead of schedule and under budget but still be a technical failure' (*idem* 1987: 120).

Bresnen refers to Gobeli and Larson and observes that the attempts to describe what a matrix organisation actually is have often been confounded by lack of agreement concerning the precise form of the parameters for defining it. A comparative description of some of the varying models and their relationship to Galbraith's original continuum is prepared by Bresnen and is shown in Figure 4.12. He notes that 'What in all cases approximated to an incipient co-ordination model of matrix management was effectively overridden by the strength of commercial consideration and the hierarchical power struggle embodied in the contract'. (*idem* 1990: 120).

Larson and Gobeli concluded that projects relying on a functional organisation or a functional matrix were less successful than those which used a balanced matrix, project matrix or project team. The project matrix outperformed the balanced matrix in meeting schedule and outperformed the project team in controlling cost.

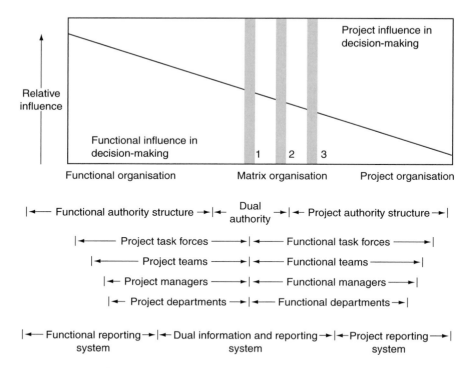

Figure 4.12 Models of matrix management
Source: Bresnen 1990
Reproduced by kind permission of Routledge

A recent study undertaken by Liu and Walker (1998) casts doubts on previous studies regarding the evaluation of project outcomes. They pose the questions below, which they proceed to answer throughout their paper:

> What constitutes satisfaction?
> Who are the claimants on the project whose feelings on satisfaction are important?
> What is the relationship between success and satisfaction?
> How should these issues inform our judgement of the outcome of construction projects?
>
> (ibid.: 209)

Liu and Walker address the problems of evaluating the effectiveness of construction project organisations and suggest that these problems result from temporary multi-organisations and from shifting multigoal coalition characteristics. They observe that the concept of project success has remained ambiguously defined and that this leads to disagreements between project participants whose own perception of the project outcomes is related to their expectation of success and the amount of effort they are willing to exert to achieve this. This concept is described as the behaviour–performance–outcome cycle (BPO). In addition they investigate the relationship between success and satisfaction and refer to Lewin (1958) who concludes that success and failure do not depend on the absolute level of performance attained but on performance in relation to one's personal standard. Liu and Walker point out that this is therefore likely to lead to differences in opinions amongst the project team as to what constitutes success.

They conclude their paper by observing that because of the variability and individuality of goal identification, definition measurement and evaluation the project critical success factors (CSFs) are likely to be highly individual and project specific and therefore that a search for generally applicable CSFs may be misplaced.

Conclusions

This chapter has examined the design of project organisation structures. It has shown that the major problem within the construction process is the temporary multi-organisational nature of the project 'team'. Identification of the need for the construction participants to understand the client's business was shown to be of significant importance.

The application of team-building concepts as a means to formulating client-focused objectives has been examined and the suggestion that team members be selected by psychometric testing was also explored. The concept of 'empowering' the individual and the project team has been proposed by Tampoe (1989), and Newcombe (1997) has noted that construction management is based on the principle that performance was shown to be achievable if the project team interacts with one another more often.

The possibility of redesigning project organisations so as to form 'work clusters' has been proposed and is currently being examined by the Tavistock Institute on two projects. Research conducted by Cox and Townsend (1997) also proposed the need for 'fit-for-purpose' relationships between contractors, subcontractors and suppliers as a means of reducing functional organisational barriers.

The work of Sidwell (1990) in showing that a project organisational structure evolves during the project lifespan and the environmental and technological variables which impinge on the project organisation, (Shirazi *et al.* 1996) have been examined. The use of systems theory as a means of examining the project's key subsystems has been proposed by Moore and Moore (1997) who suggested that this will help in designing an appropriate organisational structure for projects.

The decision and communication process during the construction phase of a project was examined; studies undertaken by Thomas *et al.* (1998) and Mead (1997) highlighted the need for accuracy and timeliness in communications and the use of IT as a means of accessing data. The cost associated with coordination and communication and the 'hierarchies' and 'markets' concepts have been introduced by Pietroforte (1997) who concluded that cooperation and information roles and rules are complementary and cirumvent the traditional hierarchical communication process.

The relationship between organisational design and a project's success was examined. The benefits of developing a matrix or project type of structure was examined, and the work of Bresnen (1990) in studying matrix forms in construction projects also provides useful insights. Research undertaken by Liu and Walker (1998) from an industrial and organisational psychology view highlights the point that a search for generally applicable critical success factors may be misplaced.

It would appear that significant changes in the way in which the construction industry operates are taking place. The impetus for these changes (client power, partnering, concurrent engineering, use of IT, etc.) may be attributed to client demands and economic pressure bearing on the industry. It is clear, however, that the combination of these 'change variables or concepts' is playing a major role in redefining the way in which the organisations participating in a project work together. It would seem appropriate to explore the concept of concurrent engineering as this process brings contractors, subcontractors and suppliers together at the design stage, with the client and design team, and must surely result in different patterns of coordination, integration and communication from those of a project procured under traditional type arrangements.

The use of the work-clusters technique and project or strategic partnering arrangements offer alternative organisational arrangements which would be expected to show different patterns of coordination, integration and communication from those projects arranged traditionally. These changes, together with the influence that the top clients have over the construction process, are leading to a 'paradigm shift' whereby project organisations can be more thoughtfully designed.

References

Ahmad, I.U. and Sem, M.K. (1997) 'Construction project teams for TQM: a factor element impact mode', *Construction Management and Economics* 15: 457–67.

Anumba, C.J., Kamara, J.M. and Eubuomwan, N.F.O. (1996) 'Encapsulating the voice of the customer in construction projects', *Proceedings of ARCOM*, 11–13 September, Sheffield Hallam University, Sheffield, 170–9.

Barlow, J. (1997) 'Institutional economics and partnering in the British construction industry', seminar paper, *AEA Conference on Construction Econometrics*, Neuchatel, February, 20–1.

Barlow, J., Jashapara, A. and Cohen, M. (1998) 'Organisation learning and inter firm partnering in the UK construction industry', *The Learning Organisation Journal* 5(2): 86–98

Belbin, R.M. (1981) *Management teams: why they succeed or fail*, London: Heinemann.

Bennett, J. (1985) *Construction project management*, Sevenoaks, Kent: Butterworth.

Boisot, M. (1983) 'Convergence revisited: the codification and diffusion of knowledge in a British and Japanese firm', *Journal of Management Studies* 20 (2): 169–90.

Boisot, M. (1988) 'The iron law of fiefs: bureaucratic failure and the problem of governance in the Chinese economic reforms', *Administrative Science Quarterly* 33*: 507–27.

Bowen, A.P. and Edwards, P.J. (1996) 'Interpersonal communication in cost planning during the building design phase', *Construction Management and Economics* 14: 395–404.

Bresnen, M. (1990) *Organising Construction: Project Organisation and Matrix Management*, Routledge.

Building (1997) 'Big changes in store at Tesco', 5 December, 14–15.

Building (1998) 'The Broadgate Legacy', 6 February, 40–45.

Burns, R. and Stalker, G.M. (1966) *The management of innovation*, London: Tavistock Publications.

Cherns, A.B. and Bryant, D.T. (1984) 'Studying the client's role in construction management', *Construction Management and Economics* 2: 177–84.

Child, J. (1972) 'Organisational structure, environment and performance: the role of strategic choice', *Sociology* 6: 2–22.

Cleland, D.I. and King, W.R. (1975) *Systems analysis and project management*, 2nd edn, New York: McGraw-Hill.

Coase, R.H. (1937) 'The nature of the firm', *Economics* 4: 386–405.

Construction Manager (1998) 'The essential accessory', 4(1): 16.

Contract Journal (1998) 'CRT's recommendations for efficiency', 18 February, 2.

Cox, A. and Townsend, M. (1997) 'Latham as half-way house: a relational competence approach to better practice in construction procurement', *Engineering Construction and Architectural Management* 4 (2): 143–68.

Crowley, L.G. and Karim, A. (1995) 'Conceptual model of partnering', *Journal of Management in Engineering* 11 (5): 33–9.

Cyert, R.M. and March, J.G. (1963) *A behavioural theory of the firm*, Englewood Cliffs, NJ: Prentice-Hall.

Davidson, C.H. (1987) 'The building team', in J.A. Wilkes *et al.* (eds) *Encyclopedia of architecture: design engineering and construction*, New York: John Wiley.

Duncan, R. (1971) 'Characteristics of organisational environmental and perceived environment uncertainty', *Administrative Science Quarterly* 16: 313–27.

Farzad, F. (1984) *An investigation into the influence of the level of development of the location of a construction project upon its duration, its cost and its use of critical path techniques of network analysis*, PhD thesis, University of Reading, Reading.

Gabriel, E. (1991) 'Teamwork – fact and fiction', *International Journal of Project Management* 9 (4): 195–8.

Gameson, R.N. (1996) 'Client–professional communication during the early stages of project development', in D.A. Langford and A. Retik (eds) *The organisation and management of construction: shaping theory and practice*, vol. 1, London: E & FN Spon, 437–46.

Gardiner, P.D. (1993) *Conflict analysis in construction project management*, PhD thesis, University of Durham, Durham.

Gray, C. and Suchocku, M.V. (1996) 'Rapid team integration to overcome the construction industry's fragmentation', in D.A. Langford and A. Retik (eds) *The organisation and management of construction: shaping theory and practice*, vol. 2, London: E & FN Spon, 629–39.

Green, S.D. (1994) 'Sociological paradigms and building procurement', in S.M. Rowlinson (ed.) *East meets west: Proceedings of CIB W92 Procurement Systems Symposium*, Department of Surveying, The University of Hong Kong, Hong Kong, December, 89–97.

Hardcastle, C. (1992) *An information model of the construction cost estimating process*, PhD thesis, Heriot-Watt University, Edinburgh.

Higgin, G. and Jessop, N. (1965) *Communications in the building industry: the report of a pilot study*, London: Tavistock Publications.

Hodge, B.J. and Johnson, H.J. (1970) *Management and organisation behaviour*, New York: John Wiley.

Hughes, W.P. (1989) 'Identifying the environment of construction projects', *Construction Management and Economics* 7: 29–40.

Jackson, M.C. (1991) *Systems methodological for the management sciences*, New York: Plenum.

Jennings, I.C. and Kenley, R. (1996) 'The social factor of project organisation' in R. Taylor (ed) *North meets South: proceedings of CIB W92 Procurement Systems Symposium*, University of Natal, Durban 239–50.

Kanter, R.M. (1983) *The change master*, New York: Simon & Schuster.

Kanter, R.M. (1989) *Mastering the challenges of strategy, management and careers in the 1990s*, New York: Simon and Schuster.

Kast, F.E. and Rosenzweig, J.E. (1985) *Organisation and management: a systems and contingency approach*, New York: McGraw-Hill.

Larson, E. and Drexter, J.A. (1997) 'Barriers to project partnering: report from the firing line', *Project Management Journal* 28 (1): 46–52.

Larson, E. W. and Gobeli, D.H. (1987) 'Matrix management: contradictions and insights', *California Management Review* XXIX (4): 126–38.

Latham, M. (1994) *Constructing the team. Joint review of procurement and contractual agreements in the United Kingdom construction industry: final report*, London: The Stationery Office.

Lawrence, P.C. and Lorsh, J.W. (1967) *Organisation and environment managing differentiation and integration*, Cambridge, MA: Graduate School of Business Administration, Harvard University.

Lewin, K. (1958) 'Psychology of success and failure', in Stacey and Demarting (eds) *Understanding human motivation*, Cleveland, OH: Howard Allen.

Liu, A.M.M. and Walker, A. (1998) 'Evaluation of project outcomes', *Construction Management and Economics*, 16: 209–19.

Luck, R.A.C. and Newcombe, R. (1996) 'The case for the integration of the project participants' activities within a construction project environment', in D.A. Langford and A. Retik (eds) *The organisation and management of construction: shaping theory and practice*, London: E & FN Spon, 458–69.

March, J.G. and Simon, H.A. (1958) *Organisations*, Oxford: Basil Blackwell.

Matthews, J.D., Tyler, A. and Thorpe, T. (1996) 'Subcontracting – the subcontractor's view', in D.A. Langford and A. Retik (eds) *The organisation and management of construction: shaping theory and practice*, vol. 2, London: E & FN Spon, 471–80.

Mead, S.P. (1997), 'Project-specific intranets for construction teams', *Project Management Journal*, **28** (3): 44–51.

Mintzberg, H. (1979) *The structuring of organisations*, Prentice-Hall, Englewood Cliffs, NJ: Prentice-Hall.

Mohsini, R.P. and Davidson C.H. (1986) 'Procurement, organisational design and building performance: a study of interfirm conflict', *Translating Research into Practice* **8**.

Moore, T.A. and Moore, D.R. (1997) 'Project management for the construction industry: an examiniation of a system approach', *Proceedings of ARCOM 13th Annual Conference*, 15–17 September, King's College, Cambridge, 301–9.

Morgan, G. (1993) *The art of creative management*, Sage.

Morris, P.W.G. (1989) 'Initiating major projects: the unperceived role of project management', *Project Management* 7(3): 180–5.

Newcombe, R. (1997) 'Procurement paths: a cultural/political perspective', in *Proceedings of ARCOM 13th Annual Conference*, 15–17 September, King's College, Cambridge, 523–34.

Peters, T. (1987) *Thriving on chaos*, London: Macmillan.

Peters, T. (1992) *Necessary disorganisation for the nanosecond nineties*, London: Macmillan.

Pietroforte, R. (1997) 'Communication and governance in the building process', *Construction Management and Economics*, **15**: 71–82.

Reve, T. and Levitt, R.E. (1984) 'Organisation and governance in construction', *Project Management*, **2**.

Rolstad, L.F. (1991) 'Project start up in tough practice', *Project Management*, **9** (1): 10–14.

Shammas-Toma, M., Seymour, D. and Clark, L. (1998) 'Obstacles to implementing total quality management in the UK construction industry', *Construction Management and Economics* **16**: 177–92.

Shirazi, B.C. (1990) *An interactionist approach in the study of organisational structuring and the decision making processes*, PhD. thesis, University of Bath, Bath.

Shirazi, B., Langford, D.A. and Rowlinson, S.M. (1996) 'Organisational structures in the construction industry', *Construction Management and Economics*, **14**: 199–212.

Sidwell, A.C. (1990) 'Project management: dynamics and performance', *Construction Management and Economics* 8: 159–78.

Sommerville, J. and Dalziel, S. (1997) 'Teambuilding: a key to innovative procurement: assessing Belbin's team-role self perception inventory', *CIB W92 Symposium on Procurement*, 20–23 May, Montreal, 711–21.

Tampoe, M. (1989) 'Project managers do not deliver projects, teams do', *Project Management Journal* 7(1): 12–17.

Tatum,C.B. (1984) 'Organising large projects: how manager's decide', *Journal of Construction Engineering and Management*, **10** (3): 346–580.

Tavistock Institute (1996/97) 'The lost world: virtual organisation in the UK building industry', 44–50.

Thomas, S.R., Tucker, R.L. and Kelly, W.R. (1998) 'Critical communications variables', *Journal of Construction Engineering and Management*, **124** (1): 58–66.

Thompson, J.D. (1967) *Organisations in Action*, New York: McGraw-Hill.

Walker, A. (1980) *A model of the design of project management structures for building clients*, PhD thesis, Liverpool Polytechnic, Liverpool.

Walker, A. and Kalinowski, M. (1994) 'An anatomy of a Hong Kong project organisation, environment and leadership', *Construction Management and Economics* **12**: 191–202.

Walker, D.T.H. (1997) 'Construction performance and traditional versus nontraditional procurement methods', *Journal of Construction Procurement* **3** (1): 42–55.

Williamson, O.E. (1979) 'Transactions cost economics: the governance of contractual relations', *Journal of Law and Economics* **22**: 223–61.

Williamson, O.E. (1981) 'The economics of organisation and the transaction cost approach', *American Journal of Sociology*, **87**: 548–57.

Williamson, O.E. (1985) *Markets and hierarchies, analysis and antitrust implications: a study in the economics of internal organisation*, New York: The Free Press.

Winch, G. (1989) 'The construction firm and the construction project: a transaction cost approach', *Construction Management and Economics* **7**: 331–49.

Woodward, J. (1961) 'Organisational characteristics and technology', in G. Cole (ed.) *Management theory and practice*, 5th edn, London: DP Publications, 82–3.

5 Organisational learning as a vehicle for improved building procurement

Derek H.T. Walker and Beverley May Lloyd-Walker

Introduction

A series of recent Australian studies into construction time performance revealed that neither the building construction companies nor the client undertook any meaningful organisational learning activities from their experience of managing building and civil engineering construction projects. Post-project evaluation studies were not required for the building and civil engineering projects but this was not the case for the process engineering projects. A post-project evaluation was expected by clients and provided as part of services procured for process engineering projects.

Organisational learning codifies experience gained by individuals and teams in a form that adds value to an organisation. Learning is an asset, comprising intellectual property, which can be reused to add competitive advantage. It is also an effective means by which innovation can be introduced to organisations. Innovation is often generated from team members' personal experiences brought with them from one temporary organisation, or teams within them, to another.

We argue that the building industry should not only initiate post-project evaluation but also enter a process of organisational learning during the entire life cycle of the project. We demonstrate how critically examining a project history can be used to help gather lessons to be learned and experiences to be shared and how these can be stored for appropriate later access. We also explore how the training and education of project participants can be enhanced.

Organisational learning and procurement systems

Clients should expect efficiency gains as a result of product innovation, changed work methods (process innovation) and the positive impact of competition derived from companies capitalising upon their organisational knowledge assets. Also, clients should expect advice that allows them to achieve more sustainable projects with lower life-cycle costs, improved quality of product delivered and quicker construction times. Clients should, therefore, expect improvements in the project delivery process.

Research into changes in construction time performance of Australian construction projects between 1979 and 1992 revealed that construction time performance productivity increased by between 19 per cent and 38 per cent. This productivity gain can be explained by technical improvements in the process of managing construction and team relationships improvement by those involved in the construction process (Walker 1994).

The construction industry has no formalised process, insisted upon by the client, to ensure that reflective learning takes place in the form of rigorous project reviews. In a series of studies (reported upon in Walker and Sidwell 1996) only in 6 projects out of 45 had formal post-project evaluation been undertaken and these 6 projects were process engineering projects. A total of 33 of the remaining projects were building construction projects, and the remaining 6 were civil engineering projects, clearly indicating that the construction industry pays little attention to finding a way to codify or record lessons learned.

A contrast was found between process engineering projects and building and civil engineering projects, which may be explained by the nature of the commissioning exercise for process engineering projects. These projects entail the commissioning of plant and equipment which usually process hazardous materials. Hence, clients and contractors feel a greater need to ensure that lessons learnt result in improvements to these projects, which are constructed for maximum safety levels when operational. Clients, such as the major oil companies, are sophisticated procurers of construction services. British Petroleum (BP) and Mobil regularly use partnering and/or strategic alliances with a critical emphasis on project partners meeting strict quality provisions for audits of the construction process in order to 'capture' lessons learned (Prokesch 1997). This does not appear to be the case in the construction industry, which has a client profile where repeat projects are less likely to be activated within several years. Mean construction time performance of the process engineering projects performed equally well compared with the other projects (Walker 1997a).

The issue of post-project evaluations not taking place poses interesting questions. For example, is the current procurement system adopted by the construction industry:

- hindering organisational learning?
- depriving clients of value from experience gained?
- depriving the construction development team (suppliers, contractors and consultants) of the opportunity to enhance their competitiveness?
- inhibiting construction development teams applying innovative processes and products?
- inhibiting sustainable development?

A further issue raised by procurement systems inhibiting longer-term improvements from organisational learning is the nature and extent of value added through organisational learning. What opportunities are being missed and what is the consequence of this? The value of organisational learning will be first addressed and linked to innovation (or lack thereof). Having established the position that organisational learning is an asset which is currently being squandered in the construction industry's traditional approach to procurement, the issue of its role in sustainability becomes important.

Through this chapter you will recognise that if clients were to embrace and stress the importance of organisational learning they would be more likely positively to influence an increase in innovation, productivity and sustainability. Organisational learning is the key to these three elements in that it provides the crucial link between theory, experience and practice. We believe that organisations that learn embrace the methodology to help them improve their performance, adapt to changed circumstances more intelligently and are better placed strategically to chart desirable futures for themselves. Clients have a role to play in encouraging organisational learning not only by encouraging all project team members to embrace organisational learning but also by requiring it as part of the procurement process in terms of deliverables. There is no reason why knowledge deliverables cannot be specified for a procurement system along with the obvious deliverable of the completed physical project.

The value of learning

What is a learning organisation and what characterises it? A learning organisation is one that encourages reflection on lessons learned, attempts to understand the dynamics of its operating environment and anticipates likely changes to that environment so as to cope with opportunities and challenges. Companies that fail fully to do so at best adapt to recent changes in order to survive but in doing so are possibly changing their context too late to be poised for the next round of challenges (Pedler *et al.* 1996). At the highest level of evolution of the learning company, according to Pedler *et al.*, the companies who reach a substantial but adaptive position in a symbiotic relationship with their environment are as likely to create their own contexts as to be created by them.

The idea of company rather than individual learning has evolved from theories of educational learning and has its basis in the work of, amongst others, Kolb (Kolb *et al.* 1971). The learning company facilitates the learning of its members and consciously transforms itself and its context. The term 'consciously' is used to impart strategic intent and the term 'context' recognises the changing world and environments that organisations face.

Organisational learning shares similar benefits to those of the planning process. Good plans seek to model possible futures in order to create a preferred future through anticipating and coping with foreseen problems or capitalising on foreseen opportunities.

Organisational learning draws upon the concept of an integrated double loop of learning between the organisation and the individual within that organisation (Figure 5.1).

By contrast in a single learning loop environment the organisation learns through integrating these two loops effectively, thereby world-views can be reframed not merely to understand the world as it appears but to anticipate how the world might be.

Figure 5.2 illustrates an energy flow model of how these four elements can be integrated. In this context 'energy' is seen as information, activity, learning, consciousness, motivation and power. Ideas feed into policy, which in turn feeds operational procedures that inform action. At the same

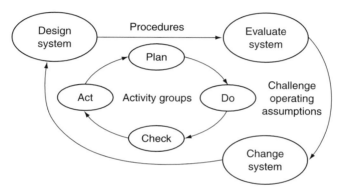

Figure 5.1 Organisational double-loop learning
Source: adapted from Argyris and Schon 1978

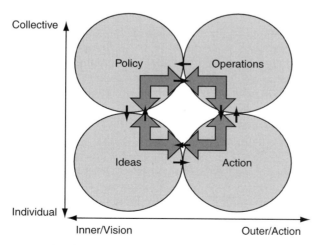

Figure 5.2 The double-loop flow of energy in an organisation
Source: adapted from Pedler *et al.* 1996

time, ideas influence action which challenges operations resulting in refurbished policy (Pedler *et al.* 1996). All this is undertaken in a reflective manner connecting individuals and organisations to facilitate improvement, adaptation and anticipation of future conditions.

Figure 5.2 also indicates that a learning organisation must be mature enough to tolerate policy, procedures and actions to be challenged. This is a constructive and creative action because challenges are directed at the way things are actually done and are not used as an exercise in rhetoric or 'power politics'. The organisational culture required for the double-loop learning model to happen effectively must be inclusive and encourage diversity. This is one of the problems facing the construction industry in taking advantage of the fruits of the learning organisation. Lenard (1996: 157), among others, has highlighted the closed and monocultural nature of the Australian construction industry in general.

Valuing diversity is necessary for innovation and productivity gains to be realised (Cope and Kalantzis 1997). Cope and Kalantzis make a strong and well-reasoned argument for valuing a wide variety of diversities to help organisations and individuals reframe their world-view that can lead to a better understanding of markets, customers, clients and colleagues. This understanding helps to unblock communication barriers, for example, to release creative energies. They argue that:

> Living in an environment where sub-cultural differences of identity and affiliation are becoming more and more significant, the third great epochal shift is to harness all that this increasing variety of community and life experience has to offer. Gender, ethnicity, religion, generation and sexual orientation are just a few words that are used to describe these differences.
>
> (ibid.: 246)

For an organisation to be able to analyse itself critically and for its members to be able to undertake reflective learning there must exist a strong tolerance of difference.

Marchand (1997) underscores the need for strategic thinking about how to improve the provision of built projects. He maintains that

> Complacency bred of past success leads to unexamined assumptions, blind spots and taboos that not only block the creation of new mandates among managers but also make it very difficult to sense, communicate and use intelligence about future trends.

It is one thing to undertake an informal review of lessons learned but quite another to build a system where learning is readily available and easily communicated to those faced with problems and issues where past learning would be of help. A structured and codified way of capturing learning is needed as no learning can take place in an organisation unless it possesses a proper memory system (Tsang 1997: 78).

This is a leadership issue. It is a leader's task to design a learning process whereby people throughout the organisation can deal with and master the critical issues they face (Senge, 1992: 345). Senge also stresses that shared vision is the product of a process that develops from extensive and often intensive communication between team members about each person's perception of what has occurred, and what is occurring. This is an organisational learning process. Through communication of different and parallel perceptions of 'history' teams can gain a truly shared vision that builds their own understanding of 'truth' (ibid.: 231). Thus shared objectives are grown and joint commitment strongly established.

Team members who take the time to share their views of events have a more valuable collective view of 'the facts' (ibid.: 235). This added value allows them to be more flexible and rapid in amending their response to changing circumstances. This may help explain why a recent study found that there was a 33 per cent improved mean construction time performance for construction management teams who exhibited a very high level of flexibility in adapting plans to meet new conditions (Walker 1996). Fisher highlighted the importance and value of organisational learning in the construction industry with an interesting insight into a Construction Industry Institute (CII) research project on a process for modelling lessons learned. She makes the point that

> Owners and contractors must depend on job end reports and/or rapid communication to transfer lessons learned from project to project. In today's fast track project environment, this is virtually impossible without a formal, systematic process that is to some extent automated.
>
> (Fisher, 1997: 36)

In a paper on organisational learning from natural disaster management, Carley and Harrald (1997) report on the activities of the Red Cross and US agencies dealing with disaster relief. They state that the Red Cross has difficulty in measuring effectiveness for post-project evaluation because they use shallow performance measures such as X meals served, and Y persons sheltered. The learning achieved identifies only what was done and not what was needed to be done. Learning must be based upon deep reflection and critical analysis, not shallow metrics. Context needs to be explored and reviewed so that past experience can be used to model new situations. This allows the additional benefit of providing a framework for simulation testing.

Constantine (1993) offers an interesting technique for use in recording organisational learning, which he refers to as 'design by walking around'. In this approach a continuous record is kept of what the group does. This is similar to what is understood as a 'global method statement' in the construction industry (Walker 1997b). Records of the process, strategy and other key management approaches together with the success or otherwise of outcomes are maintained. The recording medium is unimportant, but essentially three types of 'bins' are created for storing ideas pertaining to problematic issues. In one 'bin' success stories are recorded, noting reasons for success. In another 'bin' failures are similarly recorded; again reasons leading to this outcome are recorded. The third 'bin' contains deferred decisions, in-progress and undecided responses to issues and problems. This 'bin' can contain gems of knowledge to be used at a later time because these gems were not implemented because of current circumstances and environments not being conducive to success (Constantine 1993). This approach offers a model for undertaking continuous review and learning. In this way, organisational learning can be shared and maintained for easy access.

The main asset that a construction company has, if it does not own real estate, or much of its equipment, is its ability to carry out complex projects (Sveiby 1997: 12). This intangible asset concept is worth investigating in more detail. Sveiby (1997: 13) also draws to our attention that the value represented by management consulting firms, such as Arthur Andersen, lies in the competence of the firm's staff. The same can be said of many stakeholders and participants in a construction project. How is this asset (staff competence) to be revived, upgraded and maintained? Indeed, how is this asset developed and nurtured? If no effective attempt is made to nurture and develop this asset then firms must rely upon 'buying in' the expertise. Since the price of the service (construction or design expertise) is based in part on the cost of attracting and maintaining talented people, the client must in the end contribute to this cost. If the client contributes, then what knowledge-based deliverable provides a return on this investment? The answer is often little if any, with little or no learning being derived from the project that is formally transferred to the client. The best that can be generally offered is experience (often non-reflective in nature) derived by direct participants. Perhaps there may be a quality management system that provides documented procedures that can be of help on similar future projects. However, this is seldom the case. Additionally, as project teams tend to disband and reform with different personalities involved in the 'next' one, learning from one project is not transferred seamlessly to another. This represents squandering of assets and of scarce resources.

It is clear from the general management as well as construction management literature that the value of learning from past experience plays a crucial role in building process improvement. It is equally clear that this activity needs to be encouraged and structured into the management process. If companies are not initiating this to the extent required for significant and sustained productivity improvement, then clients should consider making the process a contractually required part of services provided by construction teams.

Innovation and organisational learning

Innovation can be defined as

> Any idea, technique and/or process, old or new, that is uniquely applied to any aspect of the production of goods and services, such that it either directly or indirectly generates measurable benefits in the form of system or process efficiency, product quality or product type.
>
> (Van de Ven 1986: in Lenard 1996)

Nam and Tatum (1997), in a study of 10 innovative projects in the USA, stress the importance of leadership by the client (owners) of the facility in fostering innovation. Some 7 out of the 10 project case studies showed owner leadership and involvement as being pivotal in adopting innovative practices. They believe that technical competence is largely responsible for this propensity for innovation. Further, their study indicates that in 8 out of the 10 cases 'managers with the authority for approval of key ideas in the construction innovation had a high level of technical competence' (ibid: 265). They also found that technologically competent managers devoted time to continued learning about technical matters. This can be achieved through undertaking post-project evaluation. They also stress the need for slack resources to be made available in the form of time or funds (ibid.: 267). Again, this can be effectively used to fund a post-project evaluation process.

Another popularised management theory of the early 1990s was re-engineering. This concept, espoused by Hammer and Champy (1994), explores the tension between what is referred to as 'inductive' and 'deductive' thinking applied to business process improvement. Inductive thinking is described as the ability first to recognise a powerful solution and then to seek the problems it might solve (ibid.:84). Thinking deductively about technology not only causes people to ignore what is really important, but also gets them excited about technologies and applications that are, in fact, trivial and unimportant (ibid.:86). Their propositions were widely regarded as a radical approach because they essentially advocated a revolutionary zeal in starting again from scratch in developing process improvement. Hilmer and Donaldson (1996), however, caution against mindlessly adopting the latest US business 'fad' or 'quick fix'. They rightly make the point that shallow thinking does not lead to productive organisational learning. While it is perfectly valid to question assumptions, they do always need to be abandoned. In fact, being innovative requires us to consider deeply how circumstances may have changed from those underpinning existing processes, and to make appropriate flexible changes to accommodate these new realities. This again requires reviewing and learning from the past, suggesting that clients who encourage this kind of review can contribute to industry improvement through organisational learning.

Interesting results arise from a study comparing innovation in the Australian construction industry with that of the Australian manufacturing industry (Lenard 1996). The sample comprised 196 survey responses from 38 manufacturers and 158 construction companies in which the adoption and use of advanced manufacturing technologies was investigated. Results showed that, 'except for systems involved in CAD/CAE [computer-aided design/computer-aided engineering], manufacturing firms used more advanced technology than construction firms, significantly so in a quarter of the comparisons' (ibid.: 103). His study reveals fascinating insights into the difference in general organisational culture between the two industry sectors and attributes much of this to the construction industry's reliance upon cost reduction for competitive advantage, with the traditional procurement system reinforcing a climate of exclusion. He reveals that:

> Manufacturers perceive their competitors as struggling to achieve a greater market share by offering innovative products or services. As such they invest considerable energy into monitoring their competitors and trying to encourage their sub-contractors to become part of their exclusive sphere of interest.

> (ibid.: 133)

The link between organisational learning and a climate of innovation is close and complimentary.

Sustainable construction principles and organisational learning

In describing an environmental management system (EMS), Hill and Bowen (1997: 236) identify as a key requirement that 'an audit report provides an essential information feedback loop to management who can take corrective action to address weaknesses of the EMS'. Several aspects of management processes are included in a 'four-pillar' EMS system advocated by Hill *et al.* (1994). Organisational learning can address sustainable construction principles through designing to meet the clients needs and objectives to minimise complexity which could demand unnecessary resources. In this way it can be seen to be in accord with constructability principles (Francis and Sidwell 1996). Sustainability seeks to eliminate unnecessary waste, including rework. In this way it can be seen to accord with re-engineering principles and waste-minimisation philosophies. All these aspects require management procedures where plans are established and where monitoring (which informs appropriate decision-making) takes place. While this process is routinely followed during the construction phase for time, cost, quality and safety, it rarely follows that a post-project evaluation is undertaken to learn from assessment of the project as a whole. Thus, a crucial opportunity for organisational learning is missed.

There is a growing awareness of the impact upon society's long-term future of the process of developing the built environment. Concepts of 'sustainable design' and 'sustainable construction', which are part of a 'green building' agenda, are becoming more important to customers and clients who wish to appear supportive of, if not actually to be practising, principles of sustainability. This requires designers and all parties to the development process to learn from past lessons and best practice principles, to promote and develop a sustainable built environment. Levin (1997) lists common features of 'green building' as practised in the USA. These include:

- energy conservation features, both passive and active in terms of energy-efficient equipment;
- water conservation features, including low consumption and use of 'grey' (recycled) water;
- low chemical emitting materials for improved air quality;
- less environmentally destructive site development to preserve natural habitats, protect excess water run-off and changes to water-tables;
- on-site waste water treatment;
- reduced ozone and fossil fuel emissions;
- life-cycle assessment approaches to material use and recycling;
- regulatory environmental impact assessment of the total building;
- regulatory recycling and waste-minimisation provisions for building materials and waste products.

(ibid.: 4)

Baldry (1997: 124) describes the contribution that project managers can make in supporting 'green building' principles. He cites the project manager's influence 'to plan a procurement and resource acquisition route which relies upon ethical and equitable commercial relationships and the pursuit of efficient resource conversion'. Interestingly, he cites the 1992 Maastricht Treaty provisions for a 'precautionary principle' relating to actions that should be undertaken to rectify environmental damage at source. He notes that raising enlightenment amongst designers and constructors to meet or exceed the Treaty's provisions is somewhat tempered by stakeholders' needs, but he indicates that there is a clear responsibility actively to monitor the development and actualisation of environmentally sustainable design and construction. Baldry identifies the project manager as one independent team member that can exert influence on design, monitoring and control to improve sustainability. His raising of ethical issues lends weight to the argument that building owners and clients are missing an opportunity to contribute to improvements to the environment in which we all share. This supports the view that organisational learning about the impact of construction upon the natural environment is important. Perhaps enlightened clients could specify in their terms of engagement that their project manager, as an independent professional, could lead the process of undertaking a post-project evaluation.

Numerous tools exist for undertaking an evaluation of various aspects of 'green building'. Graham (1997) provides a comprehensive analysis of eight tools and systems used to measure and assess various aspects of environmental impact. He concludes that these tools can be used constructively to model likely impact upon the environment for proposed design and construction strategies. Yates (1997) agrees and provides an interesting analysis of six years of use of the Building Research Establishment environmental assessment method (BREEAM) which was developed in the UK. This tool also links cost savings with environmental impact. He cites a study of 100 interviews undertaken by the Deloitte Touche Consulting Group, which reveals interesting user motivations. Interviewees comprised a range of BREEAM current and potential users who discussed reasons for using the system. General societal goals, such as ensuring that best 'green building' practices were being applied, and more client-specific goals, such as direct cost–benefit gains made from applying 'green building' principles, were cited. While Yates has a commercial interest in promoting the use of BREEAM specifically, it is interesting that large numbers of clients and building owners are concerned about 'green building' issues and are willing and keen to apply them.

Emergent issues

This chapter highlights the need for improved production through incremental productivity improvement (*kaizen*) and innovation as well as breakthrough improvements (business process improvement). Improvement in terms of sustainable construction was also discussed and it was strongly argued that there is a pressing need for practical 'green building' principles to be applied such as waste reduction, reduced rework and other benefits that can be derived from buildability or constructability principles being applied. Other environmental improvements can also be gained to profit not only the building owner or client (and indeed all stakeholders on the project production phase) but also society at large. Sustainability issues were discussed to shed light on the value of this in terms of the contribution to organisational learning and how this can be translated into improvements for users and society.

Continuing systemic change (for the better) is dependent upon building a permanent learning capacity across, as well as within, organisations (Kochan and Useem 1992:404). Although project teams tend to be temporary organisations, they are made up of individuals who form more permanent teams within their own companies. Clients also tend to have more stable organisations. It has been clearly argued that organisational learning has the capacity to raise productivity significantly. We also strongly argue that a systematic, codified and structured post-project review can reveal valuable lessons which may be successfully applied or suitably modified to improve productivity in future projects.

There needs to be an holistic approach to encouraging the learning organisation to develop. Much can be done within the organisation, as indicated above, to allow individual reflective learning as well as personal development to inform company policy so that double-loop learning can emerge, as illustrated in Figures 5.1 and 5.2. Improvement and innovation objectives can provide a vehicle for providing the momentum for this. Furthermore, external drivers need to be applied to complement and facilitate organisational learning. Incentives and enabling structures need to be put in place.

Figure 5.3 illustrates one model for encouraging the learning organisation. The company's external environment has its part to play in encouraging the learning organisation. The procurement system could provide a useful driver for change towards the learning organisation becoming the norm. Stakeholders and clients are presently loosely integrated in terms of reflective learning.

In a recent report of the Australian Industry Commission (AIC 1996) on tendering for government services it was noted that unsuccessful bidders are rarely provided with sufficient feedback after tender evaluation for them to

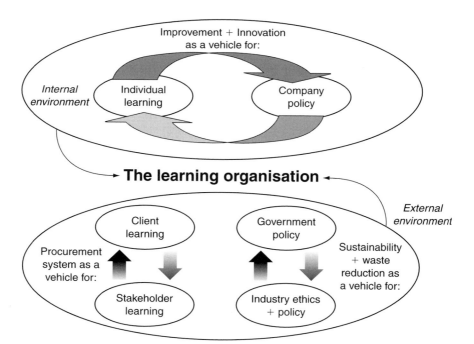

Figure 5.3 A model of how the learning organisation can be encouraged

learn effectively why they were not successful. A good example of how this can provide tangible and valuable knowledge to the client was provided, as follows:

> The benefits of debriefing unsuccessful bidders appear to be widely recognised. For instance, the Commission heard a case where an engineering firm in South Australia was able to modify the safety characteristics of future bids in the light of a safety deficiency highlighted during the post-tender debriefing – a deficiency of which the firm had been unaware.
>
> (ibid.: 346)

Lack of valuable market intelligence makes an imperfect market economy even less perfect. Improved learning by clients and those competing for contracts could improve competition to the benefit of clients and society in general.

We also suggest in this chapter that project evaluation (continuous monitoring and post-project completion) would provide a rich source of learning. Clients and stakeholders both have much to gain from this. Successful interorganisational and intraorganisational relationships rest, in part, upon sharing knowledge about stakeholders' goals and performance evaluation information so that participants understand how their goals and their level of attainment of those goals affects success of the shared project goals. Organisational knowledge about the motivation and context for the perceived motivation is important in turning plans into action as this affects the development and maintenance of trust (Howarth *et al.* 1995: 42).

Interorganisational relationships vary from co-producing subcontractors or consultants on a project through to more formal partnering or strategic alliance arrangements. The link between successful strategic direction, communication and monitoring for control has been well established for a long time; however, more recent questioning of how this is effectively achieved brings in interesting issues of stability and turbulence. Part of the perceived failure of strategic management has been attributed to an over-reliance upon stability both in the internal and in the external environments that companies confront. This is generally not the case in real life, particularly in a global economy with communication technologies that support rapid information and data exchange. This has caused turbulence which planners could neither anticipate nor perceive. The effective response is to maintain a clear vision of the strategic goals and use organisational learning to experiment and fine-tune plans or substantially to replan, depending upon what has been learned from experience (Mintzberg 1994: 209). Thus organisational learning is extremely important to the strategic planning process for all project participants.

Government and industry bodies also have played their part in attempting to facilitate greater learning. There are countless UK industry peak bodies whose charter is focused upon general support for improvement in their industry sector: the Master Builders Association and the Property Council are just two such examples. Likewise, professional bodies such as the Chartered Institute of Building and the Royal Institution of Chartered Surveyors encourage information sharing and research. The general vehicle for promoting organisational learning can be waste minimisation and sustainability policy because waste creates a burden upon society in general and purchasers in particular. If we cannot create a sustainable product or industry then our industry sectors are doomed to die.

Figure 5.3 illustrates a simplified model for using the three vehicles illustrated to drive organisational learning. Naturally, the model can be extended and developed at a more detailed level to assist in developing the learning organisation. It is important to note that in this model, communication and decision-support technology has been purposely omitted. This is not because information technology (IT) is considered unimportant – it most certainly is, as Lenard's (1996) work has clearly indicated with respect to innovation. We accept that appropriate IT strategies are vital for rapid and effective communicaiton of ideas, plans and data. We stress that for organisations to be more effective, more investment is required in human learning to be able not only to use IT effectively but also to be able to think and respond to ensure that appropriate technology is being employed. This lies at the heart of organisational learning. A useful indicative representation of the ratios of investment required by organisations is presented in Figure 5.4. The terms 'organware' and 'humanware' relate to people and their use of intelligence, and 'social software' relates to organisational and social interactions and structures to support people and their learning. This indicates clearly that there is a pressing need for more investment or a re-

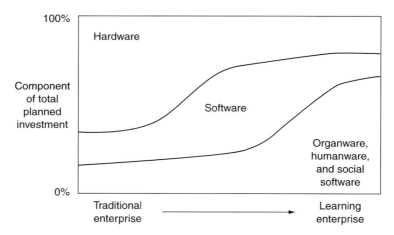

Figure 5.4 Investment associated with technology change
Source: Field and Ford 1995; adapted from Bjorn-Anderson *et al.* 1986

direction of investment from hardware towards people and how they socially interact in the workplace. While the original context of the figure was focused on IT and technological change, it captures a useful indication of the scope of the need for resource allocation needed for organisational learning.

There is a stark contrast between a shallow, unstructured and ill-conceived process for harvesting lessons learned from any project and a well-structured and rigorous examination of the 'history' of a project. On the one hand, lessons learned may be forgotten or transformed in the minds of participants well past a point of shared understanding. On the other hand, not only lessons learned but also the context surrounding events and the richness of input from numerous points of view can be harvested and put to productive use.

Post-project evaluation as an organisational learning tool

We propose that post-project evaluation is an important tool for facilitating organisational learning. It can be used inside companies, even if not required as part of the procurement system, to drive the analysis of performance to help individual and company learning for significant improvement and innovation within a company. It also, if required as part of the procurement process, can assist the client and other stakeholders in a project to learn from their experience of the project. Additionally, if a waste-minimisation element of analysis is included (in terms of physical resources, management effort and other forms of waste) then other gains can be made. In a recent study into construction materials waste, 15 per cent less waste by volume was reported, with 43 per cent less waste going to landfill (McDonald and Smithers 1998).

In terms of deliverable learning output from a project, what can a learning organisation provide that adds value? Table 5.1 illustrates the learning value that can be added to organisations.

It is important to keep in mind the context of the project when effectively charting and monitoring key performance indicators on a project to help cope with changed circumstances from those assumed at the time of preparing performance plan measures. Similarly, when undertaking a post-project evaluation a project history may again be required. It is very important to gain shared vision from the project teams during the project production phase to meet goals (Senge 1992: 207; Pedler *et al.* 1996: 56). The project history narrative helps organisations involved in these projects to understand where the project is heading or, in the case of post-project evaluation, where and how it developed from conception to completion.

If this record is well documented it can be substantially reused on other similar projects for developing method statements and plans, for development of the brief and for developing innovation based on experience. It can

Table 5.1 Deliverable learning outcomes from projects

History of the project:
This is presently fragmented in project control reports, client briefings, as-built drawings, models, etc. Greater coordination of the project history would not only assist in dispute resolution but also more importantly in learning about the context of the project and how it developed.

Lessons and experience:
Generally, individuals take this with them when they leave temporary organisations. The client takes little value from this unless the project is substantially repeated with substantially the same teams.

Training and education:
People develop from their exposure to the project. There is a role for including the education sector in this process to facilitate recognition of expertise gained in academic programmes.

also be used for dispute resolution purposes where a history of the facts surrounding the development of a dispute would be useful.

The format of the project history can take numerous forms, many elements of which are generally available as part of the project reporting system. There are generally many site-progress photographs taken on projects, and perhaps video or film recordings. There would certainly be many hard-copy records required to be maintained under current traditional procurement arrangements. These would include the time plans, cost budgeting and contract change documentation, memos and reports and other hard-copy or electronically transmitted documentation. If these were to be properly indexed and prepared for ease of access they would prove a useful resource for post-contract administration as well as organisational learning.

Lessons and experience could be distilled from project debriefing workshops. These could be undertaken throughout the project to maximise lessons learned. They could be treated as action research. If these were properly facilitated and indexed for ease of access and retrieval they could prove invaluable. Much could be learned if this were undertaken during the life of the project. Post-project evaluation workshops could also be undertaken to reflect upon experience gained of how crises were managed.

Uses in formal and informal training and education can be integrated into the proposed learning approach. Training can be enhanced through applying lessons learned, particularly when a rich source of the project context is known. Education can be encouraged through using knowledge gained from projects as the source of case-study materials to speculate on scenarios generated from case studies. Education requires greater depth of reasoning and analysis than does training.

Numerous approaches could be encouraged to use education and training to add value to the individual so that they see benefit in sharing their knowledge with the organisation, its stakeholders and the client. At the operative level, training bodies could help formally to recognise the operative's training outcomes. The impetus for academic rigour could also enhance the way in which organisational learning is developed. At the postgraduate level there is much scope for integrating organisational learning with more general dissemination of case-study research findings. Even if there is a desire for knowledge gained to be kept confidential, there should be little difficulty in maintaining confidentiality. Universities have been involved with such research for many decades and have confidentiality mechanisms in place.

Generally, there is much value that clients and project stakeholders can gain from project learning outcomes. There is little doubt that universities would be keen to cooperate, as their charter includes community service. Obviously, funding issues arise but these could no doubt be satisfactorily negotiated. The focus of this chapter is on how organisational learning can assist clients and project participants to gain added value through realising benefits from the intellectual capital of learning and so we consider detailed discussion of how universities may contribute to this is outside the scope of this chapter.

Conclusions

There are few, if any, cases where the client demands, funds and ensures that harvesting of knowledge gained from one project is available for the next. Various attempts to achieve these aims through partnering or strategic alliances are not addressing the main need – organisational learning.

A number of examples of ways in which oganisational learning can be achieved have been offered through post-project evaluation. Benefits of doing so have been explored. Organisational learning is required to be undertaken through deep thinking and analysis rather than shallow or superficial analysis of problems tackled, approaches taken, circumstances surrounding problems and solutions and strategies, or tactics, employed.

Clients have the capacity and the means to provide leadership in organisational learning through demanding well-conducted reviews of projects to record and communicate lessons learned. The cost of including this requirement is unlikely to be significant, hardly a fraction of a percentage point of project cost. Benefits to be gained from such an exercise can be lasting and significant to the client and project team members.

References

AIC (1996) *Competitive tendering and contracting by public sector agencies*, Australian Industry Commission; Melbourne: Australian Government Publishing Service.

Argyris, C. and Schon, D. (1978) *Organizational learning: a theory in action perspective*, Reading, MA: Addison-Wesley.

Baldry, D. (1997) 'The role of project management in the environmental impact of the building process', *Proceedings of the Second International Conference on Buildings and the Environment*, vol. 2, Paris 9–12 June, CIB Task Force Group 8, 121–8.

Bjorn-Anderson, N., Easson, K. and Robey, D. (1986) *Managing computer impact: an international study of management and organisations*, Norwood, NJ: Ablex.

Carley, K.M. and Harrald, J.R. (1997) 'Organisational learning under fire', *American Behavioural Scientist*, **40**(3): 310–23.

Constantine, L.L. (1993) 'Work organisation: paradigms for project management and organisation', *Communications of the Association for Computing Machinery Inc (ACM)* **36**(10): 34–9.

Cope, W. and Kalantzis, M. (1997) *Productive diversity – a new Australian model for work and management*, Annandale, NSW: Pluto Press Australia Limited.

Field, L. and Ford, W. (1995) *Managing organisational learning – from rhetoric to reality*, Melbourne: Longman Australia.

Fisher, D.J. (1997) 'The knowledge process', in L. Alarcon (ed.) *Lean Construction*, Rotterdam: AA Balkema, 33–42.

Francis, V.E. and Sidwell, A.C. (1996) *The development of constructability principles for the Australian construction industry*, Australian Construction Industry Institute, University of South Australia, Adelaide.

Graham, P.M. (1997) 'Assessing the sustainability of construction and development: an overview', *Proceedings of the Second International Conference on Buildings and the Environment*, vol. 2, Paris, France, 9–12 June, CIB Task Force Group 8, 647–56.

Hammer, M. and Champy, J. (1994) *Reengineering the corporation – a manifesto for business revolution*, Sydney: Allen & Unwin.

Hill, R.C. and Bowen, P. (1997) 'Sustainable construction: principles and a framework for attainment', *Construction Management and Economics* **15**(3): 223–39.

Hill, R.C., Bergman, J.C. and Bowen, P. (1994) 'A frame-work for the attainment of sustainable construction', *Proceedings of the First International Conference of CIB TG16 on Sustainable Construction*, Tampa, FL, 6–9 November, 13–25.

Hilmer, F.G. and Donaldson, L. (1996) *Management redeemed – Debunking the fads that undermine corporate performance*, East Roseville, NSW, Australia: Free Press.

Howarth, C.S., Gillin, M. and Bailey, J. (1995) *Strategic alliances: resource-sharing strategies for smart companies*, Melbourne: Pitman Publishing.

Kochan, T.A. and Useem, M. (1992) *Transforming organisations*, Oxford: Oxford University Press.

Kolb, D.A., Rubin, I.M. and McIntyre, J.M. (1971) *Organizational psychology*, Englewood Cliffs, NJ: Prentice-Hall.

Lenard, D.J. (1996) 'Innovation and industrial culture in the Australian construction industry: a comparative benchmarking analysis of the critical cultural indicies underpinning innovation', Faculty of Architecture, University of Newcastle upon Tyne, Newcastle upon Tyne, 406.

Levin, H. (1997) 'Systematic evaluation and assessment of building environmental performance (SEABEP)'. *Proceedings of the Second International Conference on*

Buildings and the Environment, vol. 2, Paris, France, 9–12 June, CIB Task Force Group 8, 3–10.

McDonald, B. and Smithers, M. (1998) 'Implementing a waste management plan during the construction phase of a project: a case study', *Construction Management and Economics* 16(1): 71–8.

Marchand, D. (1997) 'Looking ahead with strategic intelligence', *Business Review Weekly* 16 June, 80–2.

Mintzberg, H. (1994) *The rise and fall of strategic management*, Hemel Hempstead, Herts: Prentice-Hall.

Nam, C.H. and Tatum, C.B. (1997) 'Leaders and champions for construction innovation', *Construction Management and Economics* 15(3): 259–70.

Pedler, M., Burgoyne, J. and Boydell, T. (1996) *The learning company: a strategy for sustainable development*, London: McGraw-Hill.

Prokesch, S.E. (1997) 'Unleashing the power of learning: an interview with British Petroleum's John Browne', *Harvard Business Review* (September–October): 146–68.

Senge, P.M. (1992) *The fifth discipline – the art and practice of the learning organization*, Sydney: Random House.

Sveiby, K.E. (1997) *The new organizational wealth: managing and measuring knowledge-based assets*, San Fransisco, CA: Berrett-Koehler Publishers.

Tsang, E.W.K. (1997) 'Organisational learning and the learning organisation: a dichotomy between descriptive and prescriptive research', *Human Relations* 50(1): 73–88.

Van de Ven, A.H. (1986) 'Central problems in the management of innovation', *Management Science* 32(5): 590–607.

Walker, D.H.T. (1994) *An investigation into factors that determine building construction time performance*, PhD thesis, Royal Melbourne Institute of Technology, Victoria.

Walker, D.H.T. (1996) 'The contribution of the construction management team to good construction time performance – an Australian experience', *Journal of Construction Procurement* 2(2): 4–18.

Walker, D.H.T. (1997a) 'Construction time performance and traditional versus non-traditional procurement systems', *Journal of Construction Procurement* 3(1): 42–55.

Walker, D.H.T. (1997b) *Planning for control in the construction industry*, on CD-ROM, Melbourne: INFORMIT RMIT Press.

Walker, D.H.T. and Sidwell, A.C. (1996) *Benchmarking engineering and construction: a manual for benchmarking construction time performance*, Construction Industry Institute Australia, University of South Australia, Adelaide.

Yates, A. (1997) 'Recent advances in the assessment of the environmental impact of buildings in the UK', *Proceedings of the Second International Conference on Buildings and the Environment*, vol. 1, Paris, France, 9–12 June, CIB Task Force Grop 8, 59–66.

Part III

Emerging issues in procurement systems

6 Cultural issues

Anita Liu and Richard Fellows

Introduction

Much effort by practitioners, researchers and authors continues to be devoted to the determination of the most suitable procurement approach to be adopted for an individual project, a particular client, a project type, a client type or, even, projects in general [the idea of critical success factors (CSFs)]. Such work is linked to examination of participant and project goals (as well as various process evaluations) to arrive at conclusions concerning performance in the provision and realisation of capital or real estate projects. Further, studies are extending to incorporate requirements and realisations of ownership, occupation, use and disposal or reuse of the projects. Such holistic approaches analyse the consumption of, often non-renewable, resources embodied within the provision and use of construction projects. However, in practice, commercial imperatives – of profit, capital rationing, etc. – usually dominate perspectives of the decision-makers and hence the decisions which drive the vast majority of projects.

Built forms, most obviously in their traditional, vernacular manifestations, provide symbols of local culture or, perhaps more realistically, the local culture dominant at the time of construction. Many such symbols have disappeared, especially to remove evidence or reminders of past 'colonisation' – such as cases of the demolition of Roman buildings by the local populace on the cessation of Roman occupation – but those buildings tolerated by or even revered by subsequent populations have been preserved – such as the Forbidden City in Beijing, St Peter's in Rome, the Great Wall of China, London Bridge (now in the USA) and the Empire State Building. Even from such a small, selected sample, the likelihood of selection for preservation being dependent on the circumstances of the local society is apparent, a collective agreement or acceptance that preservation is desirable as it yields some form(s) of net benefit over other alternatives is clear and the 'highest and best use' concept of real property economics is realised through a collective view.

Hence, procurement-related decisions occur at every instance of project provision, use, adaptation, continuation of existence and disposal. Many variables impact on the decision points and decisions taken – one of the most influential and all-pervading (although, often, only implied) is culture.

From a systems perspective it is common to place culture as an environmental variable – outside the instant system's boundary but exerting influence on the system to an extent dependent upon the permeability of the boundary. Such a perspective fails to recognise that cultures are embodied in all people and therefore are present within the system and in the environment – perhaps with several diverse manifestations in both locations.

The importance of culture for management is addressed by Boyacigiller and Adler:

> Despite the spirited debate in some quarters over the influence of managerial behaviour on organisational outcomes . . . much of organisational science is based on the assumption that managerial behaviour makes a difference. The blatancy of this assumption belies the existence of deeply held cultural values.
>
> (*idem* 1991: 277)

Perspectives

Whilst culture can be regarded as the production and appreciation of arts (especially the 'high' arts), these arts are actually only particular manifestations of aspects of culture. Perspectives of what culture is range from the concept of how things are done to the definitions of culture (and, more particularly, organisational culture):

> patterns, explicit and implicit of and for behaviour acquired and transmitted by symbols, constituting the distinctive achievement of human groups, including their embodiment in artefacts; the essential core of culture consists of traditional (i.e. historically derived and selected) ideas and especially their attached values; culture systems may, on the one hand, be considered as products of action, on the other as conditioning elements of future action.
>
> (Kroeber and Kluckhohn 1952: 181)

> the collective programming of the mind which distinguishes the members of one category of people from another.
>
> (Hofstede 1980: 25)

The notion of culture is one of groups – macro and micro (generating subcultures). It is the differentiation of groups which can result in 'managerial difficulties' – such as the issues concerning construction clients and their relations with the temporary multi-organisations (TMOs) which constitute construction projects (Cherns and Bryant 1984). Thus, the nature of an individual involves consideration of character and personality [admittedly both shaped by and shaping culture(s)] whilst culture concerns collectivities.

Patterns of behaviour, artefacts and symbols may be observed and, in many cases, measured. Further manifestations of culture are heroes, icons and, perhaps most important, language (Figure 6.1). However, even such 'hard' embodiments are interpretations of a culture and are subject to interpretation by another culture. Hence, manifestations of cultures are communication devices which, consequently, are subject to the many considerations concerning effectiveness and efficiency of communications – notably, that of indexicality (Clegg 1990) in which interpretation is acknowledged to be dependent upon socialisation, including education and training.

Generally, culture is acknowledged to be rooted in people's minds – their ideas, beliefs and values. Beliefs lie at the core and become hierarchically ordered (as perceived situations demand) into a value structure which in turn underpins behaviours (Figure 6.2) thereby creating the other manifestations of culture.

In researching national cultures, Hofstede (1980) concluded that

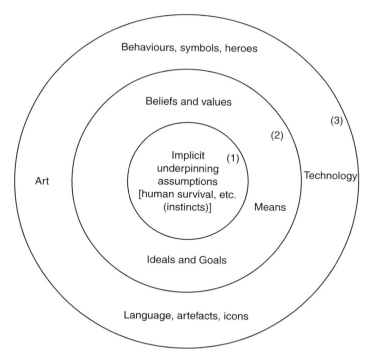

Figure 6.1 Model of cultures: levels of culture
Sources: Trompenaars 1993; Schein 1993
Notes: (1) = things taken for granted, embedded in the human subconscious; (2) = things of which people are aware; (3) = manifestations for which deciphering may be a problem

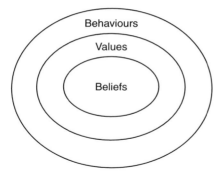

Figure 6.2 Cognitive model of culture
Source: Trompenaars 1993

appropriate dimensions were: power distance, uncertainty avoidance, masculinity–femininity and collectivism–individualism. A fifth dimension of long-termism–short-termism was added later, notably as a result of subsequent investigations in Asia (*idem* 1994).

Power distance

This dimension deals with the issue of human inequality and how within different groups the power balance is resolved. Within the work environment, this is generally evidenced in the superior–subordinate relationship. The power distance index is 'a measure of the interpersonal power or influence between [boss] and [subordinate]' (*idem* 1984: 71). Hofstede suggests that there is a socially determined equilibrium between subordinates and bosses, the subordinates seeking to reduce the size of the power distance between themselves and their bosses, their superiors seeking to maintain it.

A low value of power distance index indicates a perceived equality of ability, autonomy and independence being highly valued, whilst a high value indicates a prevalence for authoritarian behaviour, conformity and centralised decision-making.

Uncertainty avoidance

Different groups have learnt to deal with uncertainty in different ways, often because they find themselves faced with different levels of uncertainty. Adams (1965) writes of the 'risk thermostat' in relation to individuals' ability to deal with and be comfortable when exposed to risk. The principle behind uncertainty avoidance is that, at a societal level, the environment people are exposed to results in societies adopting a socially determined risk thermostat, reflected in the values of the group and the norms of behaviour, etc. which are an expression of the group's values.

Masculinity–femininity

Men and women *are* different. This is a fundamental biological fact which different societies cope with in different ways. The issue investigated here was whether this biological difference has (as opposed to should have) an effect on their roles in social activities (and hence their work values).

Collectivism–individualism

The fourth dimension looks at the relationship between the individual and the collective. Within sociology, this aspect is often referred to as gemeinschaft (low individualism) at one extreme and gesellschaft (high individualism) at the other. Within the work environment this is often explicitly linked to the individual's relationship with his or her employing organisation.

Long-termism–short-termism

The final dimension concerns the duration of perspective adopted. Typically, Westerners are concerned primarily with the instant transaction, activity or relationship and so, often pressurised by the requirements of business financiers, adopt a rather short-term view. Elsewhere, especially in Asia, people adopt a much longer-term perspective which has a major impact on how relationships are formed and conducted – the immediate transaction is not the primary focus; what is sought is an enduring relationship, whether personal or business (as in relationship marketing, seeking repeat orders, absorbing initial losses to establish an effective presence in a new market and so on).

Trompenaars (1993) advanced five value-orientational dimensions for examining cultures which, he suggested, 'greatly influence our ways of doing business and managing as well as our responses in the face of oral dilemmas' (ibid.: 29). The dimensions are:

- universalism–particularism: the relative use of rules compared with relationships;
- collectivism–individualism: the relative expression of group compared with individual identity;
- neutral–emotional: the level of expression of feelings;
- diffuse–specific: the degree of involvement;
- achievement–aspiration: the method of according status.

A number of studies (for example, Hofstede 1980) consider national cultures, where nations are delineated by politico-geographic borders. Such borders are notoriously transient and, although often convenient, may result in false generalisations. Further, sampling from a single organisation, organisation type, etc. (again, a well-known instance is Hofstede's use of IBM) may give results which are biased as a result of the sampling; commonly, organisations recruit people suited to the organisation – those who are not adequately suited soon 'quit'.

Hence, Trompenaar's (1993) perspective of culture as a normal (convenient but perhaps unverified) set of distributions is apposite – as indicated in Figure 6.3. The smaller the common culture, given similarly shaped distributions, the greater the cultural diversity is likely to be – both actual and as perceived by the other culture (often manifested in caricatures). An alternative perspective is afforded by application of the Johari window, as illustrated in Figure 6.4. The size of the 'panes' in the window may vary and change over time; on occasions, alterations may be the result of cultural changes!

Certainly, the various sets of dimensions have much in common (perhaps because many analyses have their roots in relatively similar cultures of developed Western countries). However, they do reinforce the requirement for determining the boundaries of cultural groups (communities).

A person is believed to act in ways which he or she considers to be desirable in order to give the results most satisfying to that person (the foundation of utility theory), that is, to yield most benefit according to the value structure. England (1975) suggested three primary orientations [from a personal values questionnaire (PVQ)] to predict behaviours:

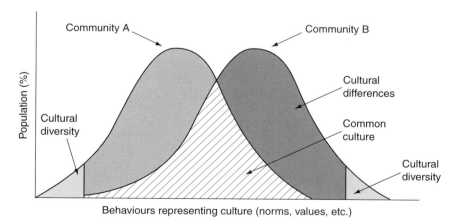

Figure 6.3 Commonality of cultures
Source: Trompenaars 1993

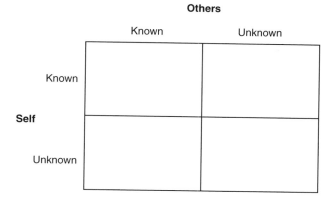

Figure 6.4 Johari window

- pragmatic: behaviour accords with concepts believed to be important and successful;
- moralistic: behaviour accords with concepts believed to be important and right;
- affective: behaviour accords with concepts believed to be important and pleasant.

Other value orientations include:

- Low-context–high-context communications (Hall 1959);
- Free Will–collectivism (Kluckhohn and Strodtbeck 1961).

People are not free to act in ways which they perceive to be most beneficial. Constraints are imposed to yield a decision environment of bounded rationality (Simon 1960) through the particular situation (time limitations, etc.) and norms of behaviour – both explicit (employer's dictates, professional institution rules of conduct, law) and implicit (moral codes, etc). Further, not only decision goals but also many of the decision constraining factors may encourage opportunistic behaviour (seeking to take 'unfair' advantage) – the presence and potency of which is likely to be a function of the prevailing culture and, as such, will impact on all situations involving governance of relationships (for example via contracts) and/or negotiations. If management may be regarded as the human activity of making and implementing decisions concerning people, appreciation of culture is central to successful management.

The term 'culture' is used increasingly commonly; in many cases it is coupled with a desire to alter a situation or process resulting in attention to cultural changes (or a change in or of culture) – almost implying, to the unwary, a set of readily available cultures with an ease of rapid inter-changeability. The majority of active or interesting management, if not all management, involves change; appreciation of the manifestations and foun-dations of cultures is powerful in determining appropriate change processes towards the desired outcomes.

Changes in peripheral behaviours (see Figure 6.2) are relatively unim-portant to those from whom change is required and so may be achieved quite readily; the more fundamental the change required, the more difficult and greater the time necessary if it is to be achieved. Changes in attitudes are acknowledged to require considerable effort over lengthy periods. Whilst changes in behaviours may be achieved rapidly, unless sustained through external means and subject to (increasing) acceptance, the change is likely to be temporary.

Thus, the formal governance of construction projects (through standard forms of contract) is usually developed reactively to changes in law and industrial practices. Contracts which have been developed and introduced to promote (normative) changes have met with great resistance and hence, generally, have been abandoned.

The nature of construction and the potential for effecting change are aptly précised by Rooke and Seymour:

> The culture of the industry has been evolved by members in order to deal `
> with its economic reality. Any attempt to change culture, which does not
> also tackle the problems which gave rise to it, stands little chance of success!
>
> (*idem* 1995)

Culture in construction procurement

Increasingly, construction procurement is considered as the combining of functions (processes and participants) to yield a product. Procurement has its commencement at the very earliest stage of a project, at conception, and may be regarded as extinguished only when the project no longer exists. However, a more pragmatic view of the period of procurement is adopted, usually from inception to handover, although facilities management (previ-ously maintenance management and life-cycle costing) is exerting pressure to extend the perspective into the project-use phase.

Commonly, performance, often in comparison with desires or targets established in the early stages of a project, is used for evaluation of the pro-curement approach employed. All too often such measures are regarded as measures of project success and of the satisfaction of participants. However, diversity of participants' goals and interests is acknowledged such that project performance may be the result of satisficing primary (powerful or influential) participants' goals and interests, including those concerned with regulations or governance of projects. The project situation is one of functional differentiation, commonly extended to fragmentation.

Given recognition that the desires, goals, etc. of the client (should) dominate the evaluation of success of a project's performance, it is helpful to consider the marketing-orientated perspective (in relationship marketing, for example Grönroos 1991). Overall performance is a combination of technical (or functional) performance and of process (or expressive) performance – what the project 'is' and how it 'works' plus the way in which it is realised (that is what is supplied – the product – and how the supply operated – the process).

Throughout project realisation, three categories of behaviour are influential – behaviours occasioned by the environment, behaviours occasioned by project organisations and behaviours of individuals. As most project participants are (members of) organisations (groups), and influential decision-makers are representatives of such organisations, consideration of organisational cultures is helpful in analysing project processes, including how behaviours may be modified and cultural changes effected.

Organisations exist both as formal and as informal groupings of people – the formal organisations include firms and professional institutions whereas the informal are, essentially, social collectives, links between the members of which range from very loose (casual acquaintances) to extremely strong (enduring, close friendships). As informal organisations develop they may become (increasingly) formalised (for example friends become family through marriage; acquaintances and friends become business partners). Often, an organisation is identified as a goal-orientated, social collective of people (Parsons 1956; Georgiou 1973).

Thus the combination of many individual organisations into a construction project TMO leads to magnification of organisational complexities which must be managed in order that good project performance and, it is hoped, participant satisfaction may be achieved.

In analysing organisational cultures, Hofstede (1994) employed six dimensions:

- Process:
    - result orientations;
    - technical and bureaucratic routines (can be diverse);
    - outcomes (tend to be homogeneous);

- Job:
    - employee orientation;
    - derives from societal culture as well as influences of founders and managers;

- Professional:
    - parochial;

- educated personnel identify with profession(s);
- people identify with employing organisation;

- open:

  - closed system;
  - ease of admitting new people, styles of internal and external communications;

- tight:

  - loose control;
  - degrees of formality, punctuality, etc. may depend on technology and rate of change(s);

- pragmatic:

  - normative;
  - concerns how to relate to the environment, notably customer's pragmatism encourages flexibility.

The alignment of Hofstede's dimensions or organisational culture with the managerial 'schools' of human-task (theory Y, theory X) orientation is quite clear; however, the development of theory Z (Ouchi 1981) provides a further perspective on the relationship of managerial style, organisational culture and performance (together with consequences). Such compatibility is reflected in Wallach's (1983) typologies of organisational culture – bureaucratic, innovative and supportive – complimenting the organisational types noted by Weber (1964).

Schein (1984) suggested two primary types of organisational culture:

- free-flowing – an unbounded, egalitarian organisation without (much) formal structure, thereby encouraging debate and (some) internal competition;
- structured – a bounded, rigid organisation with clear rules and requirements (analogous to the organic–mechanistic analysis of Burns and Stalker 1961).

Handy (1985) suggests that there are four primary forms of organisational culture:

- power, which is configured as a web with the primary power at the centre;
- role in which functions and professions provide the structural pillars for the support of the overarching top management (analogous to a Greek temple);

- task, in which jobs or projects are a primary focus yielding an organisational net (as in a matrix organisation);
- person, in which people interact and cluster relatively freely.

Handy suggests that the main factors which influence organisational culture are:

- history and ownership;
- size;
- technology;
- goals and objectives;
- environment;
- people.

Given her focus upon effecting change, Kanter (1984) suggested that organisational cultures could be examined along the dimensions of inferiority–success (as for followers and leaders amongst firms in a market) and age–youth (concerning both the individual firm and the industry: youth involves new industries, new firms, rapid and extensive innovation; age involves established, staid organisations which envisage little need for change where problems need not be confronted as they will resolve themselves).

From a distillation of literature on organisational culture, Liu and Lee (1998) identify the main orientations to be:

- power;
- rule and procedure;
- people;
- results;
- innovation;
- internal–external focus;
- team;
- customer;
- communication.

There are clear relationships between the dimensions or orientations of organisational cultures and management theories or schools. That culture has significant impacts on organisations both internally and externally has rendered recognition that its study is important, almost ubiquitous and difficult (owing to the very nature of the topic, as behavioural manifestations are rooted in beliefs), boundaries are fuzzy and interactions across them unclear. Thus, an appreciation of culture (underpinnings and manifestations) is important in all the functions of management, especially those requiring significant change.

Change involves getting from 'A' to 'B' in any and every sense; and it is everpresent as, if only because of the passing of time, no two moments are ever identical! Change is, perhaps, the only certainty. The chance of there being a constructed project (construction product) comprises a large number of individual changes involving processes and products. Hence, in procuring a project, those changes must be effected and can be achieved only through the people involved – by obtaining desired behaviours.

So, the issues of goals and motivation must be addressed in striving to secure the desired performances – indeed, researchers such as Robbins (1983) and Miner (1988) regard organisations as systems which coordinate people and other resources to achieve performance goals.

Porter *et al.* (1975: 78) believe that goals must be defined so as to focus 'the attention of individuals and groups [and] to provide a source of legitimacy for decisions'. Further, official and operative organisational goals may be distinguished (Perrow 1961), and Pervin (1989) asserts that goals are arranged in a (flexible) hierarchy.

Goals can be explicit, implicit or assumed. Given explicit goals, provided communication is effective, they are available to be employed by management; in such cases the communication process and managerial actions can be the foci of attention in examining performance against goals. Where goals are implicit (in existence and/or importance) or assumed, the determination of the goals is an added focus of attention in performance scrutiny.

Whilst many organisations do have expressed goals, as in the documents used to set up a company or in the statutory definition of a partnership, they tend to be 'establishment' goals to give the organisation as wide a scope as possible and so, in terms of operational performance, tend to be rather meaningless. Even the popular 'mission statements' tend to comprise utopian generalities with inadequate acknowledgement of constraints. Further, it is common (from anecdotal evidence from a large sample of construction industry professionals) for such organisational objectives, if they exist, to be communicated poorly to members of the organisation. Hence, organisational goals tend to be assumed, often in accordance with (bounded) rationality – such as profit-maximisation or profit-seeking – which, of course, accord with the host cultures of those having to make and act upon such assumptions.

Given that the goals of individual organisations in construction project TMOs are somewhat obscure(d) amongst decision-makers (managers) 'at the workface', the determination of goals for a project is much more problematic. A complication is that the power structure on a project is likely to alter over the phases of project realisation such that any hierarchy of goals will vary. Awareness of such a situation will encourage those who exercise power early in a project to put in place governance provisions (by fixity of design, procedure controls, contracts, etc.) to 'concrete' the goals, etc. which they have established, thereby restricting the inputs of subsequent participants. Commonly, such goal-setting actions, even if executed with the best intentions, lead to performance parameters for the subsequent participants and, as they may influence how a project is procured too, may not lead to 'best' performance (of either product or process).

For construction projects, project goals are set very rarely and communicated to even major participants less often. Of course, goal indicators are provided as milestones, budgets and prices, quality specifications, etc., likely to be determined by one or few participants, perhaps based on some information, assumptions, own desires and expediency.

The project realisation process, modelled in Figure 6.5, shows four (potentially) culturally distinct groups – clients who wish or have to use construction to pursue their own business interests (the derived demand of construction), designers and constructors who produce the projects, and regulators, of various levels of association with the construction industry. Constructors may include suppliers, etc. who see their continuity linked to construction very closely, whilst 'grand designers' often consider themselves to be distinctly separate from the 'dirty business' of construction. The traditional fragmentation of the functions necessary for project realisation has fostered differentiation between construction management, structural engineering and the other professions and functions, thereby leading to degrees of cultural distinctions. However, integration of functions – multidisciplinary design practices, design and construct organisations – plus attention to buildability and constructability and encouragement towards establishing better relations (organisational interface management) between participants (for example, see Latham 1994 on partnering) seek to overcome cultural differences (and the effects of cultural similarities, for

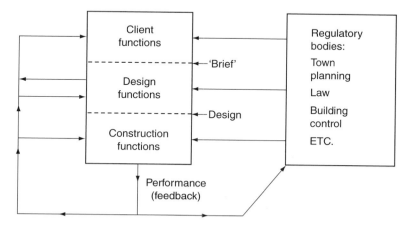

Figure 6.5 The project realisation process
Note: performance leads to satisfaction of participants and hence to (perspectives of) project success; performance–satisfaction–success also produce feedforward in the 'cycling' of project data and information to aid realisations of future projects through participants' perception–memory–recall filtering ('experiences')

example profit-maximising and opportunistic behaviours) detrimental both to project performance and to participant satisfaction.

In order to determine appropriate goals for a project, the goals, requirements and constraints of participants must be determined to achieve reconciliation within the project's cultural environment – structures available, norms of behaviour, governance provisions, regulatory requirements, etc. Given acceptability of the importances attributable to the goals and so on of the different participants, satisficing is likely to be employed to decide the hierarchy of goals for the project. Such goal determination requires sensitivity to other participants' requirements – a major consideration in negotiation and intercultural success.

Perhaps the primary factor in goal determination is elicitation of client requirements. Naive clients, who have very rare contact with the industry, generally hold a poor image of the industry and hence are cautious but may be influenced highly by the first point of contact – often an architect or other design consultant, possibly established following word-of-mouth contact with a business colleague. Such clients are unaware of what information is needed for project realisation and, indeed, of what is feasible. Unfortunately, Mackinder and Marvin (1982), in a study specifically of architects but also likely to be appropriate to all designers, found that architects are poor at obtaining briefing information necessary for project realisation.

Expert clients (that is, those who have in-house design or construction professionals, etc.) and experienced clients (that is, those who build frequently and who usually have ready access to design or construction professionals) are increasingly demanding and drive, through financial power, for changes to improve project performance (for example 30 per cent cost reduction by year 2000 for UK).

Experienced and expert clients are prepared to employ new and more analytical approaches (learning from other industries and other countries) in driving for improved project performances – for example the British Airports Authority's (BAA's) use of value management in design which, although prolonging the design phase, leads to reduced overall time and cost. Such knowledge, plus power to insist on its use, is vital to effect the change.

Lera (1982: 48) noted that 'the tradition persists whereby the architect prepares a sketch plan from which the other consultants work. Frequently alone and, often, in a matter of hours, the architect arranges spaces in a structure using predominantly aesthetic criteria'. Mackinder and Marvin (1982) further noted that, in solving design problems, architects are heavily reliant on experience. Hence it appears that in the absence of significant, effective (perceived) external pressure primary design decisions (and hence major resource commitments) are determined by aesthetics and experience and so, given the commonality of blaming rather than praising plus a more vivid recall of failures than successes, a major force to perpetuate performance *status quo* is evident.

Culture and performance

In many cultures, competition is considered to be helpful in encouraging performance in games and in business contexts. However, in business, it is the elements or criteria of competition which are being questioned – the long-held belief in price competition is being supplemented, if not replaced, by non-price considerations (notably 'quality') to yield the criterion of value maximisation. In market societies, the profit-maximisation objective is being replaced by turnover maximisation, subject to a (minimum) profit constraint (Baumol 1959), as a result of the divorcing of the ownership and management of organisations and of the recognition of the need to accommodate the goals of both in organisational performance. However, in cultures which emphasise harmony, particularly in dealings with others, a competitive (win–lose, zero–sum game) approach to negotiation not only is quite alien but also is likely to have detrimental consequences.

Higgin and co-workers (Higgin and Jessop 1965; Higgin *et al.* 1966) emphasised the value of the informal system being highly adaptive, but within the overall framework of the formal system, in the UK construction industry. Clegg (1975, 1990) investigated the operation of the adaptive system through which participants negotiate on the basis of their own perceptions of the construction process to arrive at an acceptable (usually forecast individually) outcome; in most instances, contracts (and other formal governances of business relationships) provide procedures for resolution of disputes should negotiations fail. He states (1990: 135), 'in a complex inter-organisational reality, such as a construction site, the interests embedded in different knowledges, different organisations, different hierarchies and different levels in the same hierarchies, are complex, consequently, the interpretations are rarely uncontested'; he goes on to explain the added difficulties arising from indexicality.

According to the work of Temporal (1990), the reward system is a key indicator of organisational culture. Commonly, performance-related pay is employed in construction both for operatives and for management (productivity bonus wage schemes; annual bonuses) and is related to the fact that the lowest price bid remains the factor deciding which firm 'wins' the project bidding – whether via traditional, capital price bidding or some price financial package bid [international major projects, build–operate–transfer (BOT) etc.]. Such financial emphasis is encouraged in many societies through the 'health' of shares on the stock markets being dependent upon a steady, if not growing, stream of dividend (coupled with maintenance or growth of capital investment in real terms). Such financial performance imperatives are enhanced by the high-security, short-duration investments required by banks (notably in the UK in comparison with, say, Japan). Indeed, many financial practices which also encourage ownership linkages and commonality of identity in Japan are contrary to the statutes regulating companies' activities in the UK. Thus, in the UK major, powerful financial institutions manifest both the underlying beliefs in monetary wealth but also practice short-termism to the likely detriment of (some) users of those finances.

There are significantly different views of what constitutes a 'good deal' – the outcome of a negotiation. Edgeworth box analysis in welfare economics, and similar analyses of bargains, indicates that on the basis of certain underpinning behavioural assumptions (bounded rationality for utility maximisation), within the feasible region for striking a deal, the theoretical outcome (contract curve) is likely to be distorted in practice to reflect the negotiating expertise and bargaining power of the participants. The 'get the most for me' view reflects a short-term, likely one-off, capitalist approach (win–lose) whilst the 'both or all parties leave the negotiation equally satisfied' view reflects a long-term, repeat-business approach (win–win), with a continuum of alternatives in between.

In the realm of bidding for work by design consultants and constructors there is increasing emphasis on the likelihood of achieving quality (by the client directly, and via advisors). Thus it is common for a form of two-stage selective tendering to follow in which the, say, three lowest bidders make presentations to the 'client' (using the personnel they will employ to manage the project, if successful) demonstrating how they will manage the realisation of the project. Further, it is usual for curricula vitae of key personnel to have been submitted with the other bid documents (increasingly including a priced bill of quantities, or similar price analysis).

There is much emphasis on quality assurance (QA) by clients, who are prepared to employ only those firms which have formal QA registration, which is perhaps a more widespread version of the contractor or consultant classification. More recently, attention has advanced to the use of total quality management (TQM) to amend the performance assurance sought, from getting 'what is specified' to ensuring that 'what is specified is (most) suitable'.

An additional development reflects the importance of time – in essence, the time value of money as in reduced project finance costs (for client, at least) coupled with the possibility of earlier sale of the completed project, earlier commencement of a stream of rental income or analogous benefits. As the benefit of earlier completion normally far exceeds the additional cost incurred, acceleration terms in construction contracts and/or 'negotiation' of early completion at bidding are becoming common.

Thus, through the use of tools such as value management, the utility profile of the client can be established and applied to the project to produce suitable and achievable performance goals. For such goals to secure enhanced performance they must be communicated properly and be recognised as achievable challenges by the participants concerned.

Communication plays a central role in performance. Reluctance to communicate is reflected by Hofer and Schandel (1986: 441), who state 'it is not always wise to communicate the company's plans completely or precisely to middle and lower management for various political and social reasons'. One hopes such 'strategic considerations' will not be too common or applied in other contexts!

Communication reticence may be based on fear of competition – from other firms or people – or on other fears. However, cultures, via ethics, lay down norms of acceptability – for example, how site quantity surveyors report the profitability (profile) of a project to the contractor's head office as construction progresses (feeding in early profits at later stages to achieve a smooth profit profile). Such provision of what it is assumed others want to see could have disastrous results.

Other communication issues concern indexicality and, in a more extreme form, problems of good communications between high-content and high-context language or expressions (for example English or Chinese) cultures. Without sensitivity to the receiver(s) and care in producing and transmitting the message, it is likely that clarity will be poor and/or that offence may be taken, especially if the unsuitability of messages is sustained. For either or both reasons, project performance will suffer.

Locke *et al.* (1988) note four major determinants in relating goals to performance:

- legitimate authority of the goal setter;
- peer and group pressures towards goal commitment;
- participants' expectations (that effort yields performance);
- incentives and rewards (to enhance commitment to goals).

The first two factors emphasise establishment and communication of the goals whereas the third and fourth factors emphasise realisation. Legitimate authority to set goals can occur through governance mechanisms – such as being stipulated in a contract – or by command (rather than demand) of authority and respect from other participants; the latter situation is likely to invoke more peer and group pressures (owing to the presence of respect), especially if consultation has occurred and/or the goal-setter has (obviously) considered the needs of others.

Hence, in determining goals, cultural factors concerning authority, individualism and uncertainty operate. In pursuance of goals, aspects of wealth and status apply to determine appropriate incentives or rewards to motivate desired behaviours, which requires analysis of what the particular groups, from whom the behaviours are required, value – notably the debate over whether wage or salary is a motivator or a hygiene factor!

Fortunately, the issue of safety is receiving worldwide attention not only because of its economic and financial consequences but also because of its human and social consequences; a 'culture of safety' (lack of hazards and accidents) is sought. Legislation, management actions, codes of practice, etc. are employed together with sanctions for non-compliance (stop works, fines, joint and several liability) to bring about behavioural changes generating safer working practices. Such measures (even with sanctions for non-compliance aimed at the responsible individuals, with no possibility of 'indemnification' by the employing organisation) although useful will vary in effect as they adopt the approach of a 'management tool'. The 'bravado or macho' behaviour of personnel who take risks to demonstrate something about themselves (to gain admiration of peers?) is well-known. However, policies and regulations concerning non-admission to the site of persons who have consumed alcohol have met with much success in the UK. Hence, measures to improve safety in such societies can operate through the management-tool approach.

At the other end of the spectrum lies the fatalistic view – that, irrespective of what safety measures are taken, what will happen will happen. In societies with such beliefs it is much more difficult to change behaviours and any changes are likely to require a long time to occur.

Hence, for issues such as safety, a self-determination–fate perspective seems to be helpful for the analysis of local culture and thence for determining appropriate actions to effect behavioural changes. In any event, there will be application, but to varying degrees, of the following cycle: legislation requiring behavioural changes – behavioural changes – changes accepted as (new) norms. The approach will operate only if influential members of the industry desire the changes.

A concern of much industry, including construction, is 'learning from the Japanese' which, on occasions, has meant adopting Japanese practices. Unless such adoption has been preceded by careful analysis, notably of cultural factors, the practices are unlikely to be accepted into the other society and so, for that and other reasons of inappropriateness, the desired results will not be obtained.

An example has been the Japanese achievements in terms of quality – from the cheap copies of products of some 50 (and less) years ago to the quality (and, in many cases, value-cost) leadership of many Japanese firms. The quality circles, theory Z, components of quality achievement are well-known, but the refinement, rather than innovation, practices have cultural roots (for example, language characters and tea have been refined from their ancient Chinese counterparts). Thus, Davis and Rasool (1988: 17) note that, 'If the success of the much publicised Theory Z in Japan indicates anything, it is the potential value of developing a management system internally consistent with societal norms and expectations!'

Cultural best practices

In summary, cultural best practice is to recognise cultural diversity and to behave sensitively towards the cultures encountered. The sensitivity requires awareness of assumptions made about behaviours – of one's own and of others, and their meanings – and of the possible underpinning values.

Culture is dynamic, as are its manifestations – perhaps especially language and behaviours – and, whilst any peripheral manifestations may change quite quickly and be used as managerial tools, the underlying values and beliefs evolve. Hence, change measures may be used to influence behaviours but the proactive–reactive relationship cycle (as in the development of contracts) must be acknowledged. The structural framework, of which contracts are part, impacts on formal and informal procedures – behavioural manifestations of culture – but does not necessarily affect the underpinning values and beliefs. For example, by calling an approach 'partnering' and employing some team development techniques (such as the forming–storming–norming–performing approach with selection of personnel to fulfil the team role components) can, at least in the short term, effect superficial cultural changes only.

Miller and Mintzberg (1983) found that, inherently, organisations are driven towards configuration to achieve internal consistency, to operate synergistically in their activities and to be in harmony with the prevailing economic, political and cultural realities.

The importance of culture in the management of, notably, TMOs,

Figure 6.6 Project cultural ('onion') layers
Note: boundaries are likely to vary in permeability

with particular poignancy for conglomerates such as construction organisations, especially those operating internationally, is emphasised by Sekaran (1983). He considers that culture impacts on organisations because cultural norms, values and roles are embedded in the ways in which organisations develop, their structures occur and formal and informal patterns of behaviour are adopted. The layers of influence are depicted in Figure 6.6.

References

Adams, J.S. (1965) 'Inequity in social exchange', in L. Berkowitz (ed.) *Advances in experimental social psychology*, vol. 2, New York: Academic Press, 267–99.

Baumol, W.K. (1959) *Business behaviour, value and growth*, New York: Macmillan.

Boyacigiller, N.A. and Adler, N.J. (1991) 'The parochial dinosaur: organisational science in a global science in a global context', *Academy of Management Review* 16(2): 262–90.

Burns, T. and Stalker, G.M. (1961) *The management of innovation*, 2nd edn, London: Tavistock Publications.

Cherns, A.B. and Bryant, D.T. (1984) 'Studying the client's role in construction management', *Construction Management and Economics* 2: 177–84.

Clegg, S.R. (1975) *Power, rule and domination*, London: Routledge & Kegan Paul.

Clegg, S.R. (1990) 'Contracts cause conflicts', in P. Fenn and R. Gameson (eds) *Construction conflict management and resolution*, London: E & FN Spon, 128–44.

Davis, H.J. and Rasool, S.A. (1988) 'Values research and managerial behaviour: implications for devising culturally consistent managerial styles', *Management International Review* 28(3): 11–20.

England, G.W. (1975) *The manager and his values: an international perspective*, Cambridge: MA: Ballinger.

Georgiou, P. (1973) 'The global paradigm and notes towards a counter paradigm', *Administrative Science Quarterly* 8: 291–310.

Grönroos, C. (1991) 'The marketing strategy continuum: towards a marketing concept for the 1990s', *Management Decision* 29(1): 7–13.

Hall, E.T. (1959) *The silent language*, New York: Doubleday.

Handy, C.B. (1985) *Understanding organisations*, 3rd edn, Harmondsworth, Middx: Penguin Books.

Higgin, G. and Jessop, N. (1965) *Communications in the building industry*, London: Tavistock Publications.

Higgin, G., Jessop, N., Bryant, D.T., Luckman, J. and Stinger, J. (1966) *Interdependence and uncertainty*, London: Tavistock Publications.

Hofer, C.W. and Schandel, D. (1986) *Strategy formulation: analytical concepts*, St Paul, MN: West Publishing.

Hofstede, G. (1980) *Culture's consequences: international differences in work-related values*, Beverly Hills, CA: Sage.

Hofstede, G. (1994) 'The business of international business is culture', *International Business Review* 3(1): 1–14.

Kanter, R.M. (1984) *The change masters*, London: Allen and Unwin.

Kluckhohn, F. and Strodtbeck, F.L. (1961) *Variations in value orientations*, Evanston, IL: Row Peterson.

Kroeber, A.L. and Kluckhohn, C. (1952) 'Culture: a critical review of concepts and definitions', *Peabody Museum Papers* **14**(1): 181.

Latham, M. (1994) *Constructing the team. Joint review of procurement and contractual arrangements in the United Kingdom construction industry: Final report*, London: The Stationery Office.

Lera, S. (1982) 'At the point of decision', *Building* 28 May, 47–8.

Liu, A.M.M. and Lee, S.K. (1998) *Organisational culture and job satisfaction of the real estate profession in Hong Kong*, forthcoming.

Locke, E.A., Latham, G.P. and Erez, M. (1988) 'The determinants of good commitment', *Academy of Management Review* **33**: 23–9.

Mackinder, M. and Marvin, H. (1982) 'Design decision making in architectural practice', research paper 19, Institute of Advanced Architectural Studies, University of York, York.

Miller, D. and Mintzberg, H. (1983) 'The case for configuration', in G. Morgan (ed.) *Beyond method: strategies for social research*, Beverley Hills, CA: Sage, 57–73.

Miner, J.B. (1988) *Organisational behaviour: performance and productivity*, New York: Random House.

Ouchi, W. (1981) *Theory Z: how American business can meet the Japanese challenge*, Reading, MA: Addison-Wesley.

Parsons, T. (1956) 'Suggestions for a sociological approach to the theory of organisations I and II', *Administrative Science Quarterly* **1**: 63–85, 225–39.

Perrow, C. (1961) 'The analysis of goals in complex organisations', *American Sociological Review* **26**(6): 854.

Pervin, L.A. (1989) 'Goal concepts: themes, issues and questions', in L.A. Pervin (ed.) *Goal concepts in personality and social psychology*, Hove, East Sussex: Lawrence Erlbaum Associates.

Porter, L.W., Lawler, E.E. and Hackman, J.R. (1975) *Behaviour in organisations*, New York: McGraw-Hill.

Robbins, S.R. (1983) *Organisational theory: the structures and design organisations*, Englewood Cliffs, NJ: Prentice Hall.

Rooke, J.A. and Seymour, D.E. (1995) 'The NEC and the culture of the industry: some early findings regarding possible sources of resistance to change', *Engineering Construction and Architectural Management* **2**(4): 287–305.

Schein, E.H. (1984) 'Coming to an awareness of organisational culture', *Sloan Management Review* **25**(Winter): 3–16.

Schein, E.H. (1993) 'Japanese management style and American managers', in T.D. Weinshall (ed.) *Societal culture in management*, Berlin: Walter de Gruyter.

Sekaran, U. (1983) 'Are U.S. organisational concepts and measures transferable to another culture? An empirical investigation', *Academy of Management Journal* **2**: 409–17.

Simon, H.A. (1960) *The new science of management decision*, London: Harper and Row.

Temporal, P. (1990) 'Linking management development to the corporate future – the role of the professional', *Journal of Management Development* **9**(5): 7–15.

Trompenaars, F. (1993) *Riding the waves of culture*, London: The Economist Books.

Wallach, E.J. (1983) 'Individuals and organisations: the cultural match', *Training and Development Journal*, 37(2): 29–36.

Weber, M. (1964) *The theory of social and economic organisation*, New York: The Free Press.

7 From conventionally orientated to developmentally orientated procurement systems: experiences from South Africa

Robert G. Taylor, George H.M. Norval, Bob Hindle, P.D. Rwelamila and Peter McDermott

Introduction

In Chapter 1 the conflict between the process of market liberalisation or market creation and industry-development goals was discussed. How this conflict impacts construction procurement policy and practice has been a strong theme in the work of the International Council for Building Research Studies and Documentation (CIB) Commission on procurement systems, CIB W92. Examples have been presented to CIB W92 from both the developed and the developing world, as well as from traditionally market-based economies and formerly centrally controlled economies. These issues are part of the remit of a new CIB group examining construction in developing countries [CIB Task Group (TG) 29]. Valuable lessons concerning construction procurement reform and its impact on or contribution to wider socio-economic goals can be learnt from experiences of other countries even where the socio-economic environment is different.

Traditional construction procurement systems (TCPS) have been founded on selection criteria which have mainly considered only cost, time and quality for that particular project. The focus is on the product. This chapter distinguishes these conventionally orientated procurement systems (COPS) from developmentally orientated procurement systems (DOPS). They are examined within the context of the moves towards the creation of an enabling environment in South Africa.

The chapter is structured so as to:

- review how the emphasis for construction procurement is moving from product to process and the evolution of the traditional construction procurement systems, identifying how they are inappropriate in a developmental context;
- provide a brief overview of current socio-economic priorities, specifically those which impact procurement activity;
- juxtapose the problems of procurement within a highly structured, developed environment and those of a developing environment;

- consider the operating realities of procurement models which are orientated towards community empowerment and democratic decision-making;
- present three case studies of procurement practice in specific projects which demonstrate how some developmental imperatives have been translated into practice;
- discuss the requirement for structural adjustment within the construction industry generally;
- give a brief review of current procurement policy and reform in South Africa which is attempting to build in a number of industry development objectives.

Construction procurement: product and process

Perhaps the most significant change that has occurred in construction procurement in recent years is that the needs of customers and clients are being considered as important by many of the role-players, not so much in terms of the suitability of the resultant product, which some will argue has always been considered, but in terms of the process which customers must endure in order to procure a bespoke product.

In a relatively short period of time, a process that was dominated by a single procurement system has changed to one that has a great number of alternative systems available. In keeping with this change a field of knowledge has been developed which is embodied in the term 'construction procurement systems'. It is concerned with the methods that construction customers use to procure new infrastructure, buildings and structures. It is an area of knowledge that is still developing, and this at quite a rapid pace. At the present time it is generally understood that there is no one system which can be offered as the solution in any number of procurement situations, rather, that any given project will have a unique set of requirements and conditions that can be matched to one of several procurement system options and where a 'best fit' can be found, to the point that a 'mix-and-match' situation exists which allows a unique method to be developed for any given customer or project.

However, all of the recent developments would seem to have resulted in construction procurement systems that are founded on basic economic principles that ignore the broader project environment. This suggests that construction clients are interested solely in achieving a sound return on investment, or 'value for money', in the construction production process. Perhaps this is not surprising when one considers that investment in infrastructure and buildings requires large sums of money to be expended and 'locked up' in perpetuity or for a considerable period of time (unless the customer is in the property business). But is this or should this be the only consideration?

Another recent change in construction has seen the role-players begin to accept the wider implications of construction on society and the environment. Perhaps as a result of pressure groups and public opinion, growth of interest in issues of sustainability is growing, in particular the pursuit of 'sustainable construction'. This awakening of the public conscience is forcing investors to consider the impact of their actions upon the environment and groups other than those who are intended to benefit directly from the investment. It introduces the concept of 'customer focus', something that is quite new to most participants in the construction delivery process. It suggests that they should consider objectives that are much broader than those which focus on the process itself, that is, the emphasis should be on 'external' rather than 'internal' considerations.

The evolution of the traditional construction procurement system

The traditional structure and procedures of the construction industry are steeped in the past. The construction and agriculture industries were probably the first industries to develop in most countries, particularly in Europe where the twin concepts of religion and governance developed a need for large structures and buildings in the forms of castles and cathedrals. It presaged the period when constructors were held in the same kind of esteem as the space engineers at the time of the Apollo rockets and the Moon landing programme. With the advent of the first Industrial Revolution came the need for specialisation and economic organisation. It was a period when the simple social class system, consisting of royalty and nobles at one end and peasants at the other, evolved into a more complicated class system. A working class and a middle class emerged which eventually led to a power contest between government and nobility.

This social reorganisation was mirrored by economic reorganisation and structures evolved in the form of the 'guilds' and 'professions', which were strongly influenced by the social class system. In construction it saw a class system emerge and a struggle for power between the developing professions. The day was won by architects. They attained the most dominant position at the customer–industry interface. Other built-environment professions proliferated but assumed positions which were subservient to the architecture profession. Construction personnel, be they contractors or craftspeople, were equated with the working classes and treated with indifference by customers and their professional advisors alike.

In this class-conscious mode, the system which has become known as the traditional construction procurement system (TCPS) evolved. It was, and is, a highly prescriptive system, founded on the belief that customers are ignorant of the process and must simply accept the advice offered by the professions. An attitude typical of the professions, and the second important trait, is that a profession or a member of a profession may not solicit or market his or her services to a potential client but must await a client's call. In this sense, the whole industry could be described as one which offers a design and construct service to anyone who cares to access it through an architect, who will then set into motion a process over which the client has little or no control.

This prescriptive system was the standard system for more than a century in the UK and in those countries that had evolved in a similar way or in those that had adopted the system imposed by UK and similar colonial powers.

Whilst there had always been some construction entrepreneurs who did not believe that the TCPS was in their own or their clients' best interest and offered alternative forms of construction delivery or procurement options for customers, it was only in the 1970s that real change began to be seen. Franks (1984) described how the first and second oil crises of the 1970s triggered the process of change in the TCPS. The oil crises created major economic turmoil in most industrial countries. It was the age of rampant inflation when the value of currencies was eroded by time. Investing in construction caused clients to suffer in two ways: first, they faced difficulties because the scale of the investment was considerable; second, and more significantly, they experienced losses because the procurement and delivery process is of long duration, sufficient for the impact of inflation to cause a major increase in the difference between the estimated and the final cost of any large project. The immediate demands for change were thus demands for a reduction in the duration of the delivery process. At first the professions attempted to adapt the TCPS for 'fast-track' delivery, but it would seem that this created other problems that resulted in customer dissatisfaction (Rwelamila 1996). As the gap between customer needs and industry performance widened, other players identified business opportunities, and the age of 'project management' dawned (Hindle and Rwelamila 1998).

This period of experimental change when new construction procurement systems were developed or adapted was also one which saw two periods of severe economic recessions in many countries. This occurred during the late 1970s/early 1980s and again in the early 1990s. These periods were characterised by a severe reduction in demand for construction followed by competition on a scale not seen since the great depression of the 1930s. During these periods the 'system', as Bowley (1966) termed it, failed to deliver. It was a period when the need for survival saw consultants and contractors competing with each other to find customers or to create demand through innovation and the adoption of proven business practices, such as those found in most other sectors of industry and commerce. It was also a period of unprecedented criticism of the construction process, its players and practices. Much of the criticism came from experienced customers – those who procured buildings frequently, described by Masterman (1992) as 'primary' customers. This period saw an awakening of customer awareness and diversification in construction delivery and construction procurement system development, driven both by competing industry entities and by dissatisfied customers.

Most of these changes occurred in private sector markets. The public sector – which was effectively controlled by the professions because of the fact that procurement departments and agencies were staffed mainly by architects or by engineers and quantity surveyors – was therefore 'immune' to such change. However, because the importance of the public sector as a major customer of construction declined in this period, as a result of severe economic recessionary conditions and the adoption of policies requiring a reduction in investment, an acceleration in private sector procurement development was allowed to occur. This was exacerbated by the adoption of market liberalisation or privatisation policies. As a result, many new systems became sufficiently well established for them to challenge the dominance of the traditional procurement system. This attracted the attention of public sector civil servants and politicians who began to view them as respectable and to consider their application in the public sector.

The inappropriateness of the traditional construction procurement system in the developmental context

During much the same period of time that the changes to procurement systems had been occurring a small number of construction researchers, led by Turin (1973), had concerned themselves with the suitability of the TCPS and industry structure in developing countries. Indeed, a growing number of researchers have found that the traditional approach is unsuitable for developing countries (for example, see Drewer 1975; Ofori 1980; Edmonds and Miles 1984; Wells 1986; Rwelamila 1996). A number of these workers have held the belief that capital expenditure on infrastructure and public buildings should benefit the country both by its existence and by the economic activity generated in the delivery process. Most of the poorest of the developing countries found that foreign consultants and contractors were used for construction projects as well as for the major projects in those countries that were more developed, reducing the growth potential of the local peoples and capacity building. Several of them describe the potential of construction as an engine for growth for other sectors of the economy, its potential to create jobs and its role in socio-economic development.

Wells (1986) reported her findings, resulting from case studies undertaken in Kenya, Tanzania and Cuba. Amongst the problems that she found in developing countries (though the degree would vary depending upon the level of industrial development) were as follows:

- inadequate capacity;
- lack of procedural norms;
- lack of skilled labour;
- lack of suitable materials;
- lack of plant and equipment;
- absence of support industries.

She also found that the problems inherent in the UK model of the construction industry were exacerbated in these situations.

Two recent studies are relevant here – one a pilot study from South Africa (Rwelamila and Meyer 1996) and one an extensive study from Botswana (Rwelamila 1996). There are two common factors between these countries: the inheritance of a ready-made construction framework from the UK, which is closely linked with the TCPS, and the fact that 80 per cent of the contractors and consultants in Botswana are South-African-based.

The results from both of these studies suggested that hybrids of the traditional construction procurement system (HTCPS) are basically used as 'default systems'. There was some evidence that those who were involved in the public and private sectors of both construction industries had failed to consider the issue of appropriateness. The TCPS approach seems to be adopted where no one actively considers alternative procurement options; they simply fall back on that which they have used for years without considering how it may cope with changed conditions. Overall, it appeared that the TCPS works quite satisfactorily when the power which it regulates is well distributed. However, in the cases of Botswana and South Africa, the institutional framework has been subjected to a very different set of pressures and the system has not worked effectively. The pressures in these cases have come about through radical administrative and policy changes, coupled with a need to develop domestic construction capacity rapidly (in Botswana) and to develop emerging contractors (in South Africa).

These findings in the construction sector are hardly surprising when placed in the developmental context. The transactions and activities of humankind have long been modelled on the experiences of the developed world. It has historically been assumed that norms and systems arising from a particular set of experiences in the developed world can be readily adopted by developing countries. This type of thinking is typified by Rostow's (1971) doctrine of the 'stages of economic growth', whereby the economic emergence of nations was hypothesised to be consistently and universally similar, thus ignoring national circumstances, value systems or current priorities.

In fields other than economics, developing countries have often demonstrated the inappropriateness of normative or procedural models whose evolutionary context is not their own. One of the most lasting legacies of colonial rule must be the continuation, in virtual perpetuity, of practices and procedures which may be sound in principle but which are founded on inappropriate paradigms.

However, national construction industries do share many common features. Issues of industry restructuring, globalisation of construction activity, employment creation and environmental concerns are not unique to developing countries. Uniqueness derives from national capacity to respond and the context within which that response occurs.

The rest of this chapter is about the need for the construction sector in South Africa to develop appropriate responses to the demands which it is currently required to service. There are two dimensions to this. These are roughly demarcated by the hitherto dichotomised structuring of the South African economy. First, informal contracting has become a major factor within the construction sector. As such, it requires institutional support. Second, formal contracting is bound to restructure and, indeed, has begun to do so.

Social and economic imperatives

The South African government, in a green paper, defined a broad set of socio-economic imperatives to which the policies for development of the construction industry must be responsive (DPW 1997). These imperatives include, *inter alia*, sustainable employment creation, affirmative action, the active promotion of small and micro-enterprises and the development of public sector capacity to manage the delivery process.

For the construction industry, these factors reinforce the status of process over product. In essence, the means of product delivery (or the procurement process) assumes a status which it has hitherto not been afforded. For the delivery of social infrastructure, the means towards the end of product realisation has development imperatives attached to it which are as important as the product itself. These development imperatives find expression in the organisation of projects and the methods adopted for their implementation.

The imperatives defined by the green paper are indicative of the need to reassess the set of criteria which would typically be utilised in an evaluation of procurement systems.

Procurement activity and evaluative criteria

Harvey and Ashworth (1993: 120) have defined the scope of procurement activity as being that set of activities which commences with the establishment of client requirements and objectives and ends at project completion. They add the further important observation that procurement procedures are dynamic, evolving 'to meet the changing and challenging needs of society and the circumstances under which the industry will find itself working' (ibid.). The set of activities which defines procurement may be consistently similar. The content of these activities may, however, be unrecognisably different from one situation to another. The degree of difference is likely to be intensified where developmental goals are a major project objective.

The evaluation of procurement systems, even in conventional circumstances, is not easy. Skitmore and Marsden (1988) have provided some discussion and a useful analytical framework for the evaluation of alternative procurement systems. The thinking embodied in that work is, however, underpinned by the contextually legitimate imperative of efficient, rapid product delivery. The identification of speed (of completion), certainty (of price and completion date), quality and aesthetic appeal, complexity (of the construction process), risk transfer via the contracting process and price competition are all entirely acceptable criteria for the establishment of the utility of a particular procurement system relative to client requirements. The focus, however, is on the product and the immediate issues which impact the delivery of that product.

The developing world would certainly attach high value to the efficient delivery of construction products in terms of the benefits which accrue through use and access to services and amenity. A developmentally orientated approach to procurement would, however, be one which afforded considerable prominence to those imperatives identified in the green paper (DPW 1997). This would ensure that socio-economic priorities also enjoy serious attention as part of any evaluation of the procurement process.

Procurement in developing environments can be broadly conceptualised as being concerned with the setting of fundamental development trajectories for emergent communities. System evaluation should therefore ideally occur on a basis which is uniquely developmental in its orientation and which is responsive to the specificities and resource base of the location in which development is to occur.

This section of this chapter addresses the translation of these ideals into practice, first by juxtaposing some of the main features of conventionally orientated, as distinct from developmentally orientated, projects. The discussion is then developed to include problems inherent in the implementation of developmentally orientated projects.

As part of his analysis of empowerment, Friedman (1992) has provided a useful comparative tabulation of the characteristics of developmentally orientated projects relative to conventionally orientated projects. In Table 7.1 we have adapted Friedman's original to include some elements of practice which now enjoy policy support.

Friedman's conceptualisation of empowerment via project procurement activities relies very heavily on the participation of community interests in the design and implementation of the work. Various evaluations of construction projects carried out in the Third World support the positive effects of adopting this view. For example, Materu (1986), in an evaluation of sites-and-services projects in Tanzania, concludes strongly in favour of the following success factors:

- active community participation at design and implementation stages so as to facilitate ultimate capital, as well as on-going service and cost recovery;
- the concurrent consideration of design, maintenance and cost recovery;

Table 7.1 Comparative tabulation of characteristics of developmentally orientated projects (DOPS) relative to conventionally orientated projects (COPS)

DOPS	COPS	POLICY OR PRACTICE SUPPORT MEASURE FOR DEVELOPMENT WORK
Financial assistance goes directly to the poor	Financial assistance to the state	Direct end-user subsidy scheme
They are relatively inexpensive, especially in terms of foreign requirements	They are relatively expensive in terms of foreign exchange requirements	
They are people-intensive; face-to-face interaction is essential	Capital-intensive advanced technology *and skills* are used, generally imported from abroad and displacing existing practices	Local material and labour utilisation with strong training and skills development component
Appropriate technology is used, often as an extension of existing practices		
Management is flexible (changes are *easily* possible in the course of implementation)	Management is bureaucratic	
They are fine-tuned to local conditions	They are procrustean: what does not fit must be 'cut off'	Formation of social compacts; participative design and implementation process
They are orientated toward mutual learning between external agents and local actors; transactive planning is used	Top-down technocratic planning is used; little learning occurs	
Control for negative side-effects is relatively easy and quick	Control for negative side-effects is delayed	Formation of social compact and local conflict resolution
Start-up time is short	Start-up time is long	

Source: adapted and extended from Friedman 1992

- provision of economic opportunity for project participants not only in terms of immediate job creation but also in terms of longer-term skills development and capacity growth;
- flexible management, contractual and payment procedures which are appropriate to an intensely uncertain and indecisive environment.

Operating realities associated with implementing developmental objectives

Almost inevitably, none of the above is easy. The concepts of community participation and empowerment, community contracting, sustainable intervention and skills development can be elusive ideals. The achievement of ideals can, however, be facilitated by appropriate institutional (and funding) support. Operating problems associated with community participation, community contracting and sustainability of interventions are situations of specific note. These are elaborated upon below.

Development theorists and practitioners typically place considerable emphasis on the participation of communities relative to decisions which affect their lives. Community is not, however, easily defined. This is particularly so in urban settings and is exacerbated in areas of relatively recent and rapid settlement. In the Durban metropolitan area, specifically in the location of one of the case studies described later, major settlement only began in 1985/6 and was substantially complete by 1989. The initial political circumstances of that settlement and subsequent socio-political upheaval has meant an inherent fluidity amongst residents which is not conducive to the formation of a strong and lasting communal purpose.

The coexistence of people therefore may be a necessary condition for the emergence of community identity. It is not, however, a sufficient condition. There is ample local evidence to support the view that, in the absence of immediate communal threat, there is only limited incentive for communities to behave cohesively. Rakodi (1993) attributes this to the highly individualistic nature of the survival strategies which characterise the behaviour of the poor.

There is similarly little reason for community organisational capacity to exist. An espousal of the ideal of community participation in decisions which affect communal life must be seen against the background of a typical absence of effective pre-existing structures. This adds to the task of project management. It also implies the need specifically to fund the development of appropriate capacity in order to aid the progress of development. Real participation in technical processes can occur only as a product of educational empowerment in the area of decision-making. For example, Rakodi (1993) explicitly concludes that much supposed participation is only co-option and manipulation, rarely leading to empowerment and reallocation of resources.

In South Africa, restrictions placed on funding support to include only those projects which are the result of the formation of a social compact has obligated the South African development industry to align itself with community participation as part of its normal *modus operandi*. It has also provided communities with some incentive to organise as communities, should they wish to access development funding.

The initiation of community contracting involves, in the first instance, the recreation of a work ethic in communities where unemployment has been the norm rather than the exception. Statistics for South Africa as a whole place formal unemployment at 50 per cent. About half of this number are, however, informally occupied in a variety of service-orientated occupations.

Community contracting, designed to create employment, represents a specific mix of procedural and training inputs. Labour-based methodologies are not new. Phillips *et al.* (1992) and Watermeyer (1992) report favourably on the use of labour-based contracting in differing environments. Such projects do, however, involve innovation in the fields of organisational design, management and the redefinition of the roles of consultants. In addition, documentation, price negotiation and contract letting must acknowledge the typically limited knowledge base of the recipient community.

In recent times, South African interest groups have given specific acknowledgement of the legitimacy of a labour-based approach to construction works. This was achieved through the conclusion of The Framework Agreement (SAFCEC 1993), in terms of which professional bodies have organised contracting in the engineering sector and the trade unions have established mutually protective, but also facilitative, codes of conduct for the continuance of labour-based programmes. As a document, the agreement was refreshingly realistic in that it exempted employer bodies from many aspects of potentially restrictive labour legislation, while enabling the monitoring of exploitative practices, all within the spirit of facilitating labour-based construction processes.

The Framework Agreement therefore created the necessary institutional mechanisms for giving effect to developmentally driven approaches to procurement, specifically in South African urban environments.

From a funding perspective, the favoured status which is afforded to labour-based housing and infrastructure delivery and the willingness to pay training premiums provide meaningful encouragement to the development industry in support of the achievement of developmental goals.

The goal of sustainability has many facets. For example, environmental impacts associated with projects are one such facet. Economic and affordability criteria are others. In the case of affordability criteria, the procurement process has specifically to consider community capacity to afford service charges relative to the standard of servicing which community interests and local authorities might desire. Systems which cannot be supported financially by the local community are destined to collapse.

The process of arriving at reasonable, affordable solutions has been considerably facilitated by reconsideration of local authority funding as well as by the resetting of standards. An example of this is the South African Council for Scientific and Industrial Research's (CSIR's) so-called 'Red Book' which was issued in 1994 under the auspices of the new National Housing Board. This provided institutional support for appropriate standards of intervention.

In the developing world, any procurement process which fosters or

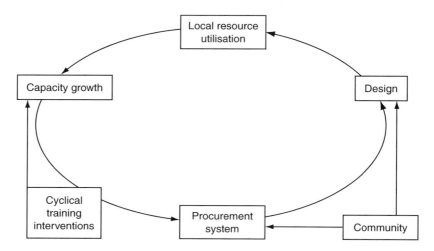

Figure 7.1 Cyclical conceptualisation of procurement response to capacity growth

reinforces dependencies must be regarded as inconsistent with the principle of sustainability as a measure of success of the process. Figure 7.1 places procurement in a systemic context whereby each project cycle is conceived to be accompanied by a growth in capacity, followed by an adaptation of the procurement process to suit that capacity growth. The figure illustrates the influence of procurement on design, which in turn influences local resource utilisation and capacity growth, all leading to a potentially revised approach to procurement. The system is shown to be sensitive to training and community intervention.

Three case studies are presented in order to illustrate the principles which have been discussed up to this point. These projects differ in location, community circumstances and scale. Common elements are, however, noticeably present in all three. Design is generally community-driven, documents and contracts are simplified and the role of project management or consultants is to train and to guide.

Case study 1: the Besters project – 8100 site informal settlement upgrade

The project is situated in the Durban metropolitan area. The situation at project inception was that of an unserviced environment, densely settled (an average of 55 units per hectare) by a community numbering approximately 48 000 persons, accommodated in informal dwellings. The area forms part of the Greater Inanda Region, which has an informally settled population of 750 000 people. The project therefore forms part of a considerably greater whole.

The project was initiated by a non-government-organisation (NGO) in 1990. The purpose of the project was to attempt to demonstrate the viability of *in situ* upgrading as a means for the delivery of housing services. Following intensive engagement of the community over a period of nine months, the project was able to commence, under the control of a community development committee. The community continued to be engaged by the NGO so as to be a meaningful participant in the design and layout of the infrastructure. Funding for the project was secured from the Independent Development Trust and the City of Durban.

From the perspective of procurement, the project was conceived as a labour-based operation whereby members of the local community would be engaged to contract for particular aspects of the work. The community formed an employment committee which was responsible for recruitment, dismissal and dispute resolution. The NGO provided the necessary educative inputs and on-site training. It also supplied materials to the operation. Productive engagement of community contractors was possible after a period of two to three weeks.

The contract documentation was reduced to two pages, from the industry standard of 15. (This approach is somewhat different from the work done in townships elsewhere in South Africa, where full documentation was used but accompanied by even lengthier documents which were designed to explain the contents of the first. The ostensible purpose behind this latter approach was to develop an early appreciation amongst aspirant contractors of the real picture. The idea is questionable for a variety of reasons, specifically connected with literacy levels. It has worked primarily because the target was somewhat more 'sophisticated' than was the case in the Besters situation.) The requirement for each contractor was that the contract be signed once. The actual work was then allocated in small packages – that is, never likely to exceed two weeks in total duration – by means of a works order. This works order formed the basis for the control of the material issues to contractors. Further works orders would not be issued until satisfactory completion of the previous ones had been achieved. The process of contract letting, including comparison with a conventional system, is depicted in Figure 7.2.

Payments were made directly into bank accounts, which the NGO opened in the name of each contractor. The payment interval was weekly, and measurement of work completed therefore had to occur at weekly intervals. The contractors were always present when the assessment was being done.

On-going training and skills development occurred during the project. This took place on two levels. The first was associated with technical skills improvement, which involved the foreperson employed by the NGO. The second was associated with entrepreneurial and management skills advancement, which was sublet to another NGO. This represented the cyclical advance of capacity which is depicted, in principle, in Figure 7.1.

176   *Robert G. Taylor* et al.

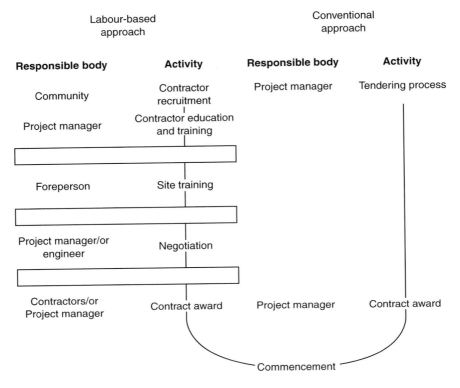

Figure 7.2 Comparative procurement procedures
Note: boxes represent points of decision for contractors to proceed or not

Materials were manufactured on-site to the fullest extent possible. All blockmaking and pre-casting was conducted as part of the project. Security of materials and their effective distribution, mostly in off-road conditions before the roads were completed, represented major initial problems for the project.

The delivery of funding to the project did not specifically facilitate the process of community engagement and training. This became something of a dilemma in terms of continuity of developmental inputs to the project. The project did, however, run on time and within budget in spite of this limitation.

Case study 2: staff housing and training block at Ndundulu, Kwa-zulu/Natal

The project is situated in central Zululand at the top of the Kkwaleni Pass between the towns of Melmoth and Eshowe. Prior to the selection of contractors and awarding of the four contracts, the professional team, consisting of an architect and a quantity surveyor, conducted a workshop which lasted some five hours. This was attended by all locally based black contractors who displayed an interest in procuring some part of the works. At the commencement of the workshop, attended by 12 local builders, each completed a comprehensive questionnaire designed to establish individual levels of skill and expertise, nature of plant owned, numbers of workers employed and general acceptability as contractors to the community.

At the workshop, the nature of the structures to be built, projected contract durations, documentation to be utilised, standards demanded by the professional team and conditions pertaining to the use of local labour were, as far as possible, clarified. Several sets of working drawings, complete with specifications, were made available for scrutiny by aspirant tenderers, and a further workshop date was set at which it was intended that motivations prepared by such tenderers be submitted. Those selected were then to be given the opportunity to submit bids for one or more of the four contracts on offer. The professional team assumed the role of project manager, responsible, in addition to its normal duties on behalf of the client, for the supply of all materials. In response to a request from the four contractors, progress payments were made on a fortnightly basis.

Undoubtedly, the most important observation to the writers was the fact that all contractors took great pains to set progress on-site at the same pace, each stage in the construction process being achieved with no more than half a day's deviation between the first and the last contractor.

All four contracts were completed within the budget and on time. No disputes were recorded and the professional team concluded that the local black builders all possessed technical skills but would probably be unable to generate growth within their organisations because of a lack of working capital and management acumen.

Case study 3: clinic building programme, Maputoland, Kwa-zulu/Natal

The project is situated in north eastern Zululand, with Mocambique and Swaziland forming the region's northern border, the Lebombo Mountains the western, the Indian Ocean the eastern, and the Mkuze and Pinda Game Reserves together with the St Lucia Nature Reserve the southern.

Approximately 13 clinics were being constructed in each ward as part of the South African government's primary health programme, and, in terms of the Independent Development Trust's instructions to appointed professional teams, local communities were as far as possible to benefit under the construction programme.

As was the case at Ndundulu, contractors, selected in each case by the local clinic community committee, were appointed to erect the 50 clinic buildings. The nature of the work to be undertaken and conditions of contract, etc., were workshopped in each of the four wards, and, without exception, it was found that builders were without working capital and the contracts were of necessity to be on a 'labour-only' basis.

The absence of working capital again resulted in the project manager having to assume responsibility for the timely delivery of materials. As the clinic sites in approximately 60 per cent of the area are accessible only with the use of four-wheel-drive vehicles it proved necessary to arrange to have bulk orders delivered to adjoining Jozini, which is located on a tarmacadam road, to be distributed by light four-wheel-drive commercial vehicles. The remoteness of sites, together with the Independent Development Trust's objective of achieving community benefit, resulted in building blocks being manufactured on-site in most localities. Localised manufacture naturally placed an additional burden of quality control on the principal agent.

Difficulties were encountered with the supply of certain strategic materials, one such being crushed stone for use in concrete foundations, surface beds and wall beams. One crushing plant exists in adjoining Jozini, and the supplier does not undertake any deliveries. This resulted in a 180 km delivery distance to some of the more remote sites in the region. This journey had to be undertaken by tractor and trailer.

One material that proved problematic in approximately a third of the localities was water. The local authority was unable to deliver water by tanker to these sites because of the lack of adequate access roads, and local housewives were employed to carry water from streams, often some 5 km distant, accomplishing two trips per day and delivering 20 l per trip. With the value placed on water in these regions, incidences of theft with respect to containers of water left unattended on sites overnight had to be combatted with the use of community nominated persons acting as guards.

Here the important observation to the writers is the fact that, in areas where the local contractor was obviously ill-equipped to carry out the task, the community invariably insisted upon that person being employed. It was more important that the payments remained in the area than the achievement of a good finish. Contractors proceeded at a leisurely pace, and as the inclusion of penalty clauses in contracts was strongly opposed by both parties to such agreements, completion dates were entered to be exceeded.

Lessons from the case studies

The use of the abundantly available, and substantially unemployed, labour in urban settings in South Africa, coupled with the emergent contingent of small black-labour-only contractors who are the products of either recession or projects such as that of Besters, represent a valuable resource to the formal construction industry. They also represent something of a challenge. This challenge has been readily recognised by the Building Industries Federation of South Africa (BIFSA).

South Africa is advantaged in the sense of having a highly sophisticated formal contracting sector. However, during the economic recession which was at its most intense during the late 1980s and early 1990s the industry became depleted of the resources which it urgently required in order to meet the challenges of the Reconstruction and Development Programme of the government. This dilemma of the industry was given detailed attention in recent research completed by Merrifield (1994). This research has identified the strengthening of labour-only subcontracting as a product of recession. The productive utilisation, and stimulation, of the capacity of such labour-only subcontractors, the entrenchment of the roles of such contractors and the attendant operating, legal and institutional reforms were identified as central issues which were currently occupying the attention of industry. The specific parameters of job creation and empowerment which the government wishes to see deliberately pursued as part of procurement activity lends incentive to industry to seek effective, appropriate structural arrangements for the continuance of operations at suddenly heightened levels of activity.

The construction industry is therefore bound to restructure and adjust procurement practice accordingly, so as to align formal sector interests with those of the less formal. Scarce management expertise which resides in contracting organisations will be required, and indeed already is required, to contract on the basis of training and empowerment imperatives. New alliances will be required in order to meet the challenges which exist. Emergent relationships between mainstream contractors, NGOs and training institutions around projects are interesting, spontaneous developments. New procedural arrangements on contracting and price negotiation are part of this process. South Africa is undoubtedly involved in several fundamental paradigm shifts. Construction procurement is not exempt.

New roles and opportunities in rural areas

Procurement systems in the rural areas of Kwa-Zulu are almost without exception designed to take cognisance of parameters that have become manifest during the past decade. Foremost among these is the need for community empowerment, and the client base is thus usually representative of the population peculiar to the area. The committees representative of such communities are usually inexpert in the management functions their electorate expect of them, and such committees are often supplemented by the apointment of a suitable professional consultant.

Community representation invariably extends to the design process, and details calculated to maximise the use of local labour become integrated in the design development from inception.

The incursion of large, and usually white-owned, contracting entities operating from urban bases is generally viewed with scepticism by the rural people. Current incentives demand that procurement systems be tailor-made to suit local black rural operators, utilising as much labour as possible from the labour pool of the area where development is taking place.

The vast majority of work is awarded on the basis of labour-only input by the contractor, or a major down payment is made at inception of the works, often resulting in the disappearance of the builder before completion is achieved. The principal agent therefore invariably is a project manager responsible for, in addition to the usual duties associated with such an appointment, the ordering and timely delivery of all materials.

Of a sample of 65 rural-based contractors, only 12 purported to be capable of computing their labour-only bid by conducting measurement of individual elements. The remainder operated on the basis of the application of a metre-square rate or a calculated guess. On closer examination, it transpired that the same metre-square rate applied throughout, regardless of the shape, the floor-to-ceiling height, concentrations of internal walling, nature of services, etc.

The lack of working capital, an almost universal condition among rural black builders in Kwa-Zulu/Natal, has to date not impacted upon practitioners to any noticeable extent, as this sector of the construction industry has accounted for a very small segment of total building turnover in the region. However, the green paper may open opportunities for professional consultants to become involved with rural appointments.

Existing contract documentation is inappropriate in a rural environment, and those consultants who have ventured into this field have found it necessary to draw up papers, tailor-made for each venture. There is an apparent resistance to the application of retention funds, not because of the resentment of the concept *per se* but more because the financial impact of such an application would adversely affect a contractor with limited working capital and with modest profit and overhead mark-up. It was found necessary to transfer onus for insurances to the client. To avoid the embarrassment of defects in structures without the comfort of a retention fund at the conclusion of the works, the consultant responsible for quality control has to inspect and accept or condemn throughout the construction process. The economic need to involve the local community as far as possible in the development process also places new-found and additional onus on professional consultants and, in particular, on the project manager.

Political and social peculiarity considerations are part and parcel of the procurement process in Kwa-Zulu/Natal, and an understanding of local hierarchy and protocol is essential to the consultant. To operate in an area it is necessary to seek audience with and be introduced to the *Nkosi*, or chief, of the area, after which the formality is repeated with the *Induna*, or regional supervisor.

South Africa faces a significant development challenge both in urban areas and in rural areas. In order to meet that challenge in an appropriate manner the construction industry can no longer operate exclusively within the paradigms of the developed world. To succeed in delivery and to develop capacity the construction sector must effectively mobilise and develop all its construction resources. This requires a new paradigm to suit South Africa's developing urban and rural communities.

Post-script: procurement reform and practices in South Africa

The challenges detailed are being faced by the Government of National Unity as part of its policy development across the whole spectrum of governance. It is, in effect, a re-engineering exercise that has allowed the policy development engineers to work from a set of fundamental principles and to integrate them into all kinds of policies.

Public sector procurement, through the agency of the Department of Public Works, is one such area where these principles are being integrated into policies towards the procurement and management of public sector infrastructure and facilities. They have recognised that the organs of state have enormous collective buying power and that public sector procurement can be used as a tool to achieve socio-economic objectives. In essence, they aim to stimulate economic growth by economically empowering previously marginalised sectors of society. However, in combining a number of development objectives they have created a situation that has exposed them to criticism by those who are used to measuring only the immediate economic benefits of infrastructure delivery and development. Indeed they have experienced a degree of criticism about the overall cost to taxpayers (Rakodi 1993) and the difficulties experienced by the construction industry in adapting to the new systems.

In the early stages of policy development a 10-point plan towards procurement reform was proposed in which several ways were demonstrated that other objectives could be achieved in tandem with the construction delivery process. The plan contains the following key components:

- removal of barriers to entry by improving access to tendering information, simplification of documents and procedures and the establishment of procurement advice centres;
- breakout procurement (unbundling large projects into smaller packages for small and micro enterprises) employing local people and providing them with sustainable skills.

Many of these components have formed part of construction procurement practice, on an experimental basis, since mid-1995. A variety of procurement systems have been used, including design and build, but variants of the traditional system still predominate. Whichever system is used it will incorporate several of the requirements of the 10-point plan, though the various requirements may simply be incorporated into the contractor selection system or into the conditions of contract. Although there have been some problems in the implementation, and policy is still in the formulation stage, it appears that the principles have been accepted by the industry and that the government's objectives are being achieved, at least in the short term.

In addition to the above the procurement agency has recognised that the government lacks sufficient funds to provide the level of infrastructure required in a specific timeframe. As a result they have adopted other procurement systems which allow partnering between government and industry, using systems such as build–operate–transfer (BOT) and build–own–operate–transfer (BOOT), and these systems have been adapted to deliver in ways which facilitate the principles described above.

The South African example is one where a client has boldly stated that the construction industry must deliver that which its customers require and not that which it would like to provide, as is made clear in the following quotation:

> The construction industry exists to serve its clients and client needs must be met by the industry.
>
> (DPW 1997: 46)

References

Bowley, M. (1966) *The British building industry: four studies in response and resistance to change*, Cambridge: Cambridge University Press.

CSIR Red Book (1994) Guidelines for the provision of engineering services and amenities in residential township development, issued by the Department of Housing in collaboration with the National Housing Board, Published by CSIR Division of Building Technology, Pretoria, South Africa.

DPW (1997) 'Creating an enabling environment for reconstruction, growth, and development in the construction industry', green paper, a government policy initiative coordinated by the Department of Public Works (DPW), South African Government, November.

Drewer, S.P. (1975) 'The construction industry in developing countries: a framework for planning', Building Economics Research Unit, University College London, London, April.

Edmunds, G.A. and Miles, D.W.J. (1984) *Foundations for change: aspects of the construction in developing countries*, London: Intermediate Technology Publications.

Franks, J. (1984) *Building procurement systems*, Ascot, Berks: Chartered Institute of Building.

Friedman, J. (1992) *Empowerment – the politics of alternative development*, Oxford: Basil Blackwell.

Harvey, R.C. and Ashworth, A. (1993) *The construction industry of Great Britain*, Newnes: Butterworth-Heinemann.

Hindle, R.D. and Rwelamila, P.D. (1998) 'Resistance to change: architectural education in a turbulent environment', *Engineering Construction and Architectural Management Journal*, forthcoming.

Masterman, J.W.E. (1992) *Building procurement systems*, London: E & FN Spon.

Materu (1986) 'Sites and services projects in Tanzania – a case study of implementation', *Third World Planning Review* 8(2): 121–37.

Merrifield (1994) 'The performance and capacity of the construction industry in the early 1990s', unpublished report, National Housing Forum, Box 1115, Johannesburg, South Africa.

Ofori, G. (1980) *The construction industries of developing countries: the applicability of existing theories and strategies for their improvements and lessons for the future: the case of Ghana*, unpublished PhD thesis, Bartlett School of Architecture and Planning, University College, London.

Phillips, S., Meyer, D. and McCutcheon, R. (1992) 'Employment creation, poverty alleviation and the provision of infrastructure', *Urban Forum* 3(2): 81–114.

Rakodi, C. (1993) 'Planning for whom?', in N. Devas and C. Rakodi (eds) *Managing fast growing cities*, New York: Longman Scientific and Technical, 207–35.

Rostow, W.W. (1971) *The stages of economic growth: a non-communist manifesto*, Cambridge: Cambridge University Press.

Rwelamila, P.D. (1996) *Quality management in the public building construction process*, unpublished PhD thesis, Department of Construction Economics and Management, University of Cape Town, Cape Town.

Rwelamila, P.D. and Meyer, C. (1996) *Procurement and balancing project parameters in Botswana and South Africa*, Department of Construction Economics and Management, University of Cape Town, Cape Town.

SAFCEC (1993) 'The framework for public works projects using labour intensive construction systems', South African Federation of Civil Engineering Contractors, Bedfordview, Johannesburg, South Africa.

Skitmore, R.M. and Marsden, D.E. (1988) 'Which procurement system? Towards a universal procurement selection technique', *Construction Management and Economics* 6(1): 71–89.

Turin, D.A. (1973) *The construction industry: its economic significance and its role in development*, part 1, 2nd edn, BERU, University College, London, September.

Watermeyer, R.B. (1992) 'Contractor development in labour-based construction', unpublished paper.

Wells, J. (1986) *The construction industry in developing countries: alternative strategies for development*, Andover, Hants: Croom Helm.

8 Use of World Wide Web technologies and procurement process implications

Derek H.T. Walker and Steve Rowlinson

Introduction

The concept of using the world wide web for procuring services is gaining momentum. Construction procurement requires a complex arrangement of skills and capabilities throughout the project delivery process that may make much use of the web in the near future.

It is interesting to reflect just how quickly some technologies can make a significant and unexpected contribution to competitive advantage; the mobile phone and fax machines are cases in point. This chapter explores the potential impact of the web and reports on a survey of the perceptions of 17 senior Australian construction professionals to their current exposure to the web and on a second series of surveys on the use of information technology (IT) in Hong Kong construction companies, with reference to similar surveys in the UK and Australia. The surveys provide interesting insights into how the web is perceived and how quickly web developments in other industry sectors may be implemented to change the way that project and construction professionals respond to procurement needs in the construction industry.

The chapter then continues this theme by exploring the likely impacts of IT on procurement both in general and in global settings. Many new developments have taken place in the aerospace and automotive industries which are of great significance to the construction industry. Both of these industries have made extensive use of computer-aided design (CAD) systems in the past but are now moving towards the use of visualisation or virtual reality (VR). This technology goes beyond the conventional two-dimensional drawing representation and provides for all members of the project team a three-dimensional view of the project, be it a Boeing 777, a new Chrysler motor car or a production facility such as a factory. By moving away from reliance on two-dimensional drawing representation it is possible for virtual teams to be put together which can interactively, in cyber space, produce new designs. These designs can be manipulated or viewed by all project participants and the process of production can be considered and planned at the same time as the design is being developed. Indeed, the construction process can itself be visualised in a four-dimensional setting, time being the fourth dimension. Such technologies open up the prospects for real concurrent engineering to take place in the design and construction processes and, by allowing participants to become team members without physically moving locations, a whole range of new and vastly quicker, and perhaps cheaper, procurement options present themselves.

This chapter is structured as follows: an introduction to the technologies investigated is provided to establish definitions and provide a common understanding of technical terms used. The survey methodology is explained and then the results are presented. Discussion of results is provided and conclusions drawn. The lessons learned are then generalised, and additional ideas and directions are identified in order to postulate likely scenarios for the impact of IT on procurement in general and, in particular, in the global construction market.

The Australian survey

A survey of 17 key construction industry decision-makers was undertaken during mid-1997 to gain a better understanding of their perceptions of their awareness and the potential utility of IT communication technologies. The survey investigation of these technologies was confined to electronic mail, electronic transfer of files, voice mail, video conferencing, the Internet and the intranet. Results and analysis of this survey are presented and comments provided.

The survey provides a snapshot view based on a selected small sample of respondents. Although the survey does not comprise a representative sample of construction professionals, it does provide a useful pilot study that can be used to formulate a series of hypotheses. This study forms the basis to develop later a more rigorous international survey of the state of readiness of construction industry professionals to take full advantage of IT technologies associated with the web; hence its place in this book is that it looks to investigate future best practice. Frequently, new technologies are adopted with a remarkable speed which inhibits the ability of researchers to study the adoption and impact of such technologies. The introduction of fax machines and mobile phones in the construction industry are good examples of this.

Definition and explanation of terms

- *information technology (IT)*: generally assumed in this publication to be information and data which are delivered by electronic means.
- *Electronic mail (e-mail)*: transfer of information in electronic (digitised) form from one computer to another using special software that allows this communication to take place. The medium of communication is generally text-based and/or data files but often includes digitised images or digitised sound files.

- *Electronic transfer of documents or files or electronic data interchange*: digitised information is communicated electronically in a similar way to e-mail attachments. Electronic data interchange (EDI) provides for structured data to be exchanged electronically between users. Invoice data, for example, can be transmitted and used for billing and stock-control purposes. Most of this kind of data exchange used at present in the construction industry is the electronic transmittal of computer-aided drawing (CAD) files.
- *Voice mail*: digitally (rather than analogue) transmitted sound files that communicate voice messages. These files can be treated as any other data file; for example, they may be embedded in or attached to text documents.
- *Video conferencing*: video images can be electronically transmitted from one receiver to another. Special purpose video conferencing venues have been in use for many years. The transmission links individuals through real-time video pictures. Some software packages also have enabled users to access interactively shared software applications such as data bases, word processing or scheduling programs. More recently, however, Internet applications such as CUSeeMe® have enabled inexpensive video image transfer to be possible using interactive shared software programs. This allows for 'virtual' conferencing and problem-solving meetings to be conducted electronically, with participants in different physical locations taking part.
- *Internet communications*: the vast collection of interconnected networks that all use the transmission control protocol/Internet protocol (TCP/IP) that evolved from the ARPANET of the late 1960s and early 1970s. The Internet connects hundreds of thousands of independent networks and many millions of users into a vast global 'internet' (Surfas and Chandler 1996).
- *Intranet communications*: a private network inside a company or organisation that uses the same kinds of software that one would find on the public Internet but only for internal use. As the Internet has become more popular many of the tools used on the Internet are being used in private networks. Many companies, for example, have web servers that are available only to employees, restricted by various levels of password control. Note that an intranet may not actually be an internet – it may simply use a computer network.

The survey methodology

The literature on the construction industry's adoption of web technologies is limited and in its infancy. Much of the available information contained in articles and papers is targeted at a general construction industry audience, with a promotional message being the general tenor. While this serves to inform readers of technology potential it provides scant evidence of reflection on actual implementation. This lack of reliable information posed a barrier to the production of a reliable survey instrument that can be used to gain representative insights into the current application of web technologies in the construction industry. It was therefore decided to develop a pilot survey to test awareness of a small group of individuals, who are in an active position to influence their company, as to their perceptions of web technologies. It was also expected that the pilot survey would expose issues and questions that could be better formulated into a wide-scale survey that could be used to benchmark web technology adoption in the industry and provide guidance on best-practice case studies.

As the study was of necessity limited it was decided to select key decision-makers from a narrow range of construction industry professionals. To this end three groups of construction professionals were invited to participate. These were: contractors known to be open to innovative IT; project management consultants; partners and/or senior partners of quantity surveying practices. These three groups were considered to deal with more general information processing issues than design consultants who tend to focus on IT applications associated with design rather than administration, control and procurement systems. Eight project management consultants, four quantity surveyors and five construction managers or contractors offered to participate in the survey.

The survey questions

The first set of questions, in section A, tested the awareness of survey respondents of six types of IT (Figure 8.1). Section B tested the respondents' current level of use, the area of use and the perceived expected level and place of use over the next two to three years (Figure 8.2). Section C examined the respondents' views on expected potential use of the specified IT over the next two to three years and their view of the likelihood of its use. A total of 11 potential uses of web technologies for project management and project procurement applications were investigated (Figure 8.3).

At the end of each question, space was provided for respondents to add their own comments.

Survey results

Table 8.1 records the results of the median response value for the separate groups and the whole sample of 17 respondents. From this table a moderate overall current level of use and awareness of all e-mail, file transfer, intranets and internets is indicated. The overall group made some use of voice mail and video conferencing. The contractors claimed a higher use of all technologies than did the project managers and quantity surveyors, except for the case of intranets. The project managers seem to make less use of IT than do the other two groups. Open-ended comments provided by

Electronic Communication Activities Perceptions

Name ____

Title/Position _______________________________________

Company ____

Address ____

____

Phone #__________________ Fax # ________________

E-mail ____

A Awareness of Possible Electronic Communication Activities

Please indicate your current state of awareness of the following electronic communication technologies.

Where: 1 = very unaware, 2 = somewhat unaware (heard/read about it), 3 = aware (participated in conversations/tried some of these activities), 4 = moderately aware (occasionally use it as receiver or generator), 5 = very aware (use it as a matter of course frequently).

❒ Electronic mail

❒ Electronic transfer of documents/files/EDI

❒ Voice mail

❒ Video-conferencing

❒ Intranet Communications

❒ Internet Communications

Comments

Figure 8.1 Authors' survey, section A

B For Answers of Moderately Aware or Greater – Perceptions of Usefulness

Please state your degree of current use and expected use over the next 2–3 years.

Where: 1 = no use, 2 = little use, 3 = some use, 4 = moderate use, 5 = regular or constant use. Also where predominantly used: Office [O], At Home (Private) [P], Entertainment [E], Office to/from Home [OH].

Current Level of Use & Where Used		Next 2–3 Years Expected Use & Where	
☐ Electronic mail	Used ☐	☐ Electronic mail	Used ☐
☐ Electronic transfer of documents/files/EDI	Used ☐	☐ Electronic transfer of documents/files/EDI	Used ☐
☐ Voice mail	Used ☐	☐ Voice mail	Used ☐
☐ Video-conferencing	Used ☐	☐ Video-conferencing	Used ☐
☐ Intranet Communications	Used ☐	☐ Intranet Communications	Used ☐
☐ Internet Communications	Used ☐	☐ Internet Communications	Used ☐

Comments

Figure 8.2 Authors' survey, section B

Discussion Questions:

In considering what kinds of communications could be maintained electronically for easy access (including text, diagrams, video, voice, and image) either by select people or by a wider group, what potential lies for greater inter-firm and intra-firm connectivity over the next 2–3 years?

Where Potential Use: 1 = very low, 2 = some use, 3 = moderate, 4 = high, 5 = very high.
Likelihood of use: 1 = very low, 2 = some use, 3 = moderate, 4 = high, 5 = very high.

Potential Use Next 2–3 Years	Likelihood Of Use
❏ Project control meetings (progress)	❏
❏ Project monitoring & review	❏
❏ Estimating & planning information	❏
❏ Transfer of documents – drawings, sketches, transmittals, correspondence etc	❏
❏ Ad hoc problem solving meetings	❏
❏ Client & business presentations	❏
❏ Marketing/Sales	❏
❏ Recruitment and procurement (sub-contractors & suppliers)	❏
❏ Posting notices and/or memos to team members	❏
❏ Maintaining contact information phone, fax, etc	❏
❏ Internal/external magazine or bulletins	❏

Comments

Figure 8.3 Authors' survey, section C

Table 8.1 Awareness of information technology (IT), by sample subgroup: median for all respondents ($n = 17$), project management respondents (PM; $n = 8$), quantity surveying respondents (QS; $n = 4$) and construction management consultants (CM; $n = 5$)

IT	All	PM	QS	CM
Electronic mail	4	4	4	5
File transfer/electronic data interchange	4	3	3.5	5
Voice mail	3	2.5	3.5	4
Video conferencing	3	2.5	2	4
Intranet	4	4	2.5	2
Internet	4	3.5	3	5

Note: 1 = very unaware, 2 = somewhat aware (heard/read about it), 3 = aware (participated in conversations/tried some of these activities), 4 = moderately aware (occasionally use it as receiver or generator), 5 = very aware (use it as a matter of course frequently)

them indicated that project managers saw themselves as being more closely involved with face-to-face communications. Issues of trust building through personal contact were said to be very important by this group. The contractors commented that they saw immediate benefits from use of these technologies for routine and process-orientated communication. All groups commented that video conferencing lacked the intimacy required for trust and relationship building.

Table 8.2 provides interesting insights into current use patterns of these technologies. The usage patterns were consistent with awareness of the technology, which is to be expected. The interesting aspect of the data is that there is a clear intention to make greater use of these technologies. With the exception of video conferencing and voice mail the web

Table 8.2 Patterns of use of information technology (IT), by sample subgroup: median for all respondents ($n = 17$), project management respondents (PM; $n = 8$), quantity surveying respondents (QS; $n = 4$) and construction management consultants (CM; $n = 5$)

IT	Current use				Use over next 2–3 years			
	all	PM	QS	CM	all	PM	QS	CM
Electronic mail	4	3.5	4.5	4	5	5	5	5
File transfer/electronic data interchange	4	3	3.5	5	5	5	5	5
Voice mail	2	1.5	3	5	4	4	4	5
Video conferencing	1	1	1	2	3	2	3	3
Intranet	2	2.5	2	1	4.5	5	4	5
Internet	3.5	2.5	4	3.5	5	4	5	5

Note: 1 = no use, 2 = little use, 3 = some use, 4 = moderate use, 5 = regular or constant use

technologies are expected to be used extensively. There were some interesting comments made by respondents about their low levels of appreciation for potential advantages of voice mail. Considerable irritation was expressed concerning the inability to reach people on a one-to-one and direct basis of communication. All groups appear to favour direct interaction and resent any filters that come in their way to being able to phone a person and speak to them. Message banks, recorded messages and paging services are seen as barriers to communication and trust building. This has important implications for procurement systems, both in their operation and the choice of procurement system to use.

None of the respondents expressed an awareness of some of the benefits to filtering messages. Modern voice-mail telephony systems can allow (provided that the caller can be electronically identified) automatic filtering and/or rerouting of calls based on a preference file. For example, an organisation's central PABX can have a preference file that identifies that calls from person X should go automatically to person Y even when Y has rerouted their phone from their office to, say, a meeting room (Pullar-Strecker 1997).

Also, respondents did not see any significant advantages to holding *ad hoc* problem-solving virtual meetings by means of internet-based video conferencing systems. However, an example of this would be where technical problems arise on-site where design and construction or project management team members are widely dispersed. The ability to meet quickly in virtual space and solve the problem and, at the same time through use of shared interactive software, be able to document and action decisions reached would seem to be a highly effective use of this kind of technology. Again, such instances of technology use will become common in the next few years and will become essential elements of effective procurement systems.

It remains to be seen how voice and video transmission using web technologies is effectively used in other industries and how quickly news of such best practice cases using these technologies reaches the construction industry.

Survey participants appear generally to see a high potential for web IT, with a typically high expectation of these being used over the next two to three years (Table 8.3). Project management consultants seem slightly more sanguine than do the other two groups. Planning and communication activities are given a higher assessment of potential web IT use than estimating activities or marketing activities. Comments made were consistent with the result and highlighted an expected continued need for face-to-face contact and relationship building and maintenance as the most effective means of marketing. Quantity surveyors were slightly less sanguine than the other groups, although otherwise following a similar rationale for their opinions. The contractor group felt that there was a very high potential for web IT to be used for communication within and between teams. They also viewed the opportunities to be high for marketing and recruitment, though they were moderately confident that these technologies would be used for company

Table 8.3 Expected potential use of information technology (IT) over the next two to three years, and the likelihood of its use, by sample subgroup: median for all respondents (*n* = 17), project management respondents (PM; *n* = 8), quantity surveying respondents (QS; *n* = 4) and construction management respondents (CM; *n* = 5)

Potential for interfirm and Intrafirm connectivity through IT	Next 2–3 years				Likely use			
	all	PM	QS	CM	all	PM	QS	CM
Project control meetings (progress)	4	3	4.5	3.5	3.5	2	4	3
Project monitoring and review	4	3	4.5	4	4	2	4	4
Estimating and planning information	4	3	3.5	3	3	2.5	4	4
Document transfer	5	4	4.5	4.5	4.5	4	5	5
Ad hoc problem-solving meetings	3	3	2.5	2.5	4	3	2	4
Client and business presentations	3	3	2.5	3	2.5	2	3	3
Marketing and sales	4	4	3.5	3.5	4.5	3.5	5	5
Recruitment and procument	4	3	4	4	3	1.5	4.5	3
Posting notices or memos	5	5	4.5	4.5	5	5	5	5
Maintaining contact by phone/fax, etc	5	5	4.5	4.5	5	4.5	5	5
Internal or external magazines and bulletins	4	4	4	4	4	3.5	5	5

Note: 1 = very low, 2 = some use, 3 = moderate, 4 = high, to 5 = very high

presentations. A key technology which has the potential to change procurement systems completely is the concept of virtual collaboration, where designs may be posted and manipulated in cyber space by the collaborating groups of designers, consultants and contractors. Such systems have the potential to transform the procurement process, especially when viewed from the perspective that technology such as four-dimensional visualisation of the construction process has the potential to present the complete virtual facility in cyber space over time. Such technologies are already in place in the aerospace and automobile industries, with Dassault's CATIA system being the industry leader. Such systems offer the opportunity to integrate the whole design and construction process, even to the level of negating the necessity for detail drawings on-site, which can be replaced with a video produced from a visualisation of the construction process, for use by site workers.

Future trends

It is interesting to note that during 1997 several of the respondents made moves to increase their access to web IT. Most of the companies have a very limited access, with only one or two personal computers connected to the web. Two of the contractors are establishing intranets with perceived benefits of improved communications capability and an expectation that this will provide them with a quality competitive edge. In all cases where general comments were given there was an expectation that some cost saving could be gained, generally through improved efficiency and waste minimisation associated with quicker communication turnaround.

Issues of concern expressed by respondents relating to barriers to effective use of these technologies centred on software and hardware support and staff training and change management. This is an important issue as e-mailed messages are not effective if those receiving them do not turn on their computers or read their e-mail regularly. This is in fact a cultural issue in that the culture of the organisation, or temporary organisation such as the project team, has to change dramatically, and such change is always difficult. It ultimately involves the redefinition of roles and responsibilities and so is likely to be a major driving force or brake in procurement-system change. For change to be effective much emphasis will need to be placed on IT education and continuing professional development.

The lack of enthusiasm for video conferencing was surprising. There were some interesting comments made about the expense and general inconvenience of current arrangements of booking facilities. Some respondents are aware of less expensive and more readily available web-based video conferencing and mild interest was expressed regarding its potential use in the near future. Consultants and contractors still view their team interactions as being largely composed of social interaction and communication. The human touch, body language and the 'smell' of subtextual issues are seen as highly important in the construction industry. Such issues have already been discussed in Chapter 6 and their significance has been highlighted in the current procurement process; this significance will not necessarily be diminished in the procurement process where IT plays a much bigger role.

The most interesting insights related to the level of concern expressed by the survey respondents pertaining to the preparedness of the industry to respond to opportunities offered by use of web technologies. There were understandable concerns expressed about security of data and the potential for unwelcomed 'computer hackers' breaking into their computer systems. Much of this concern was not so much directed at the stealing of data but more on industrial sabotage. This concern is consistent with a recent survey, undertaken by the web consulting company *www.consult*, of chief executives or IT managers of medium-sized to large corporations, reported on in the *Business Review Weekly*. According to the report, 40 per cent of the managers surveyed were concerned about poor perceived security of their computer installations as a result of their web connection (Banaghan 1997).

Other concerns raised related to the level of training and support required to use web IT effectively or indeed to use any newly introduced technology or to incorporate change into an organisation. Frequently, companies commit insufficient funds to training and support when introducing computer systems. Reliance on technology was seen as disempowering, in that there is a great deal of frustration when 'things go wrong' and nobody seems to know how to deal with the problem. Most users of IT have probably experienced acute frustration when finding that files will not open properly or machines fail to function. Additionally, problems relating to training extend to lack of time to train as well as the quality or extent of training needed. Commonly, comments were expressed that everyone is already very busy with their job and has little time for extensive training in web technologies even if there were sufficient funds available to do so. The cost of training and support was also seen as discouraging in an industry with very tight profit margins.

The more gloomy comments about difficulties in adopting web IT were countered by a sense of inevitability of clients demanding their use, particularly when these technologies are used as a matter of course by some clients in the project design and production phases. It was substantially for this reason that at least two of the contractors had put in place an intranet during the latter part of 1997, and two other contractors were considering and planning the use of an intranet during 1998. Generally, there was a lot of positive support for a future where web IT would be routinely used. The principal qualifier expressed by most respondents was that the construction industry is very much a 'people' industry and that human factors such as trust and face-to-face communication are essential for effective productivity.

It will be interesting to compare and contrast results from this pilot study with a longitudinal study as well as a study of other industries where web IT is already in widespread use.

Use of information technology in Hong Kong

This section presents the results of current research into the extent that construction IT is used by the contractors of the construction industry of Hong Kong and is based on the paper presented by Futcher and Rowlinson (1998) at CIB W78 (Information Technology for Construction) in Stockholm. Futcher and Thorpe (1998) describe how the first distributed computing technologies were introduced into the public works organisation in an *ad hoc* manner in the early part of the 1980s. The computing expertise within the public works departments was low, but none the less they were responsible for specifying the functional requirements for the distributed computing systems within their organisation. As a consequence, from the outset, there has been a reliance on easy-to-use proprietary software solutions or on utility packages for the development of simple user-applications. This culture of making do with low-technology skills fostered a reliance on packaged solutions. There is now considerably more IT expertise within the public works departments and they are specifying more sophisticated information management systems, but they continue their practice of innovation using proven technologies. Futcher and Thorpe describe the recent experience of the works departments of the Hong Kong Government Special Administrative Region (HKGSAR) of a better way for construction information to be produced, and shared, for the purposes of construction project management. In the cases cited, the professionals participating in these public works projects were new to the technology but successfully worked within IT for the management of the ambitious construction projects. Although the requirement for an IT system to aid in the project management of these large construction projects merely sought to automate manual methods, an approach was adopted that involved core process design, ergonomics and the management of change. Documents are generated, recorded and stored within a document management system to enable the facts of the matter on any issue to be discovered quickly, in an interactive manner, from the desk of the manager.

The future

The HKGSAR and its public works departments are moving ahead with their use of IT, but the direction of future innovation includes the clients, consultants and the contractors involved in the projects. The question is: 'are the consultants and the contractors involved in the public works ready for this phase of IT innovation? ' The following section focuses on one aspect of current research into this subject, a postal survey of the exploitation of IT by the public works contractors of the HKGSAR. It sets out to examine the extent that IT is deployed within the Hong Kong construction industry by public works contractors, and its results are generalised to consider how IT will affect the procurement system demands of public sector clients.

The method developed by Betts and Shafagi (1997) was included as part 4 of the postal survey questionnaire. Their self-assessment health check consists of 28 questions that companies answer to assess their use and management of IT. The questions are grouped into three categories:

- the position of IT within the competitive business strategy of the organisation;
- the overall role of IT within the organisation;
- the current IT strategy within the company.

This part of the questionnaire is useful to the companies in the survey population sample because it gives them instant feedback on the extent of their strategic use of IT compared with a benchmark standard developed in the UK for this purpose. It is also useful as it provides a basis for comparison with the same research exercise carried out in Great Britain by Betts and Shafagi and in Australia by Stewart (1997).

Comparison with results from a survey in the UK and in Australia

The survey in the UK involved 11 companies whose staff participated in interviews to build a consensus for the response in the health-check questionnaire. A similar approach was taken by Stewart in his survey of 48 companies in Australia. The approach taken in Hong Kong was different. The population sample was 316 and more varied than were the UK and Australian populations. A postal survey was used in Hong Kong to obtain a response to the health-check questionnaire. The extent of consensus building that took place within each company as they decided on an answer to each of the questions is not known. The comparative results, in statistical terms, are shown in Table 8.4.

Survey conclusions

A side-by-side comparison of the overall results from each survey is shown in Table 8.5. The statistics have been converted back into the grading used in the questionnaire, except for the standard deviations. To introduce more sensitivity into the differences, the grading is further given positive or negative bias to indicate better intermediate positions (for example, B+, C−).

Table 8.4 Comparison of scores from the information technology (IT) Health-check survey in Hong Kong, the UK and Australia

	High	Low	Median	Mode	Mean	Standard deviation
Hong Kong postal survey of 84 contractors:						
Competition and business strategy	25	8	17	16	17	2.98
Role of IT	19	6	10	10	10	2.36
IT strategy	53	16	33	30	34	6.05
Total	97	30	61	51	61	11.89
UK survey of 11 Companies:						
Competition and business strategy	26	15	19	19	19	3.13
Role of IT	17	8	12	14	12	2.6
IT strategy	55	37	47	37	47	6.21
Total	96	60	77.55	84	78	11.07
Australian survey of 36 companies:						
Competition and business strategy	55	23	39	35	40	8.28
Role of IT	17	7	12	12	12	2.56
IT strategy	25	13	19	17	19	2.94
Total	94	47	70	66	71	12.05

Table 8.5 Comparison of the surveys carried out in Hong Kong, the UK and Australia

Survey of 316 Hong Kong works contractors	A	C−	B−	C+	B−	11.89
Survey of 11 UK construction firms	A	B−	B+	BA	B+	11.07
Survey of 36 Australian construction firms	A	D	B	B	B	12.05

Note: A = aware of and implementing information technology (IT) fully; B = aware of and implementing IT in part; C = aware of but not implementing IT; D = unaware of IT; a score slightly above or below a grade is indicated by a plus or minus sign, respectively; a score falling on the midpoint between grades is given the combined grade (for instance, BA)

Also, a score falling on the midpoint between gradings is given the combined grade (for example, BA).

The variability of the results obtained is similar in each of the national surveys even though there are differences in survey methodology. In the Hong Kong survey there is slight evidence that those contractors who have a quality assurance certification adopt IT for strategic reasons to a greater extent than do those without the certification. The results overall also indicate that the Group C Hong Kong contractors, who are approved for public works projects in excess of HK$50 million, perform moderately better in the assessment than do Group B or Group C contractors. The latter two groups are similar in their performance. In terms of the national surveys the results are similar: the overall mode and the overall mean indicate that the UK is performing marginally better than Australia, and that Australia is performing marginally better than the Hong Kong contractors. The strengths and weaknesses of the results from the Hong Kong contractors is shown in Table 8.6.

In terms of each of the 28 questions, the Hong Kong contractors performed 'well' overall in 64 per cent of the questions, and performed 'poorly' overall in 34 per cent of the questions. Stewart (1997) lists those areas in the questionnaire that show the Australian companies to perform 'well' and those that indicate that they perform 'poorly'. He also cites the areas in which the UK companies performed 'well' and in which they performed 'poorly'. The results for the Hong Kong survey are described by a categorisation stated by Futcher and Rowlinson (1998):

A Aware and implementing IT fully;
B Aware and implementing IT in part;
C Aware but not implementing IT;
D Unaware of IT.

Table 8.6 is a list of the paraphrased questions and the overall grading achieved by the Hong Kong contractors according to this measurement of performance. The overall performance is neither outstandingly 'good' nor

Table 8.6 Graded overall performance of the postal survey of 84 Hong Kong contractors regarding information technology (IT)

Question subject	Unaware	Aware but not using IT	Aware and using IT in part	Aware and committed to IT development
IT support of core competencies			Yes	
How IT helps the company compete			Yes	
Impact of IT on goals and objectives			Yes	
Position of IT compared with other technologies			Yes	
Impact of IT on clients			Yes	
Belief that IT expertise wins work			Yes	
Use of IT as part of strategic alliances			Yes	
Current use of IT in the company			Yes	
Relationship between IT and business strategy			Yes	
Participation of IT in forming business strategy		Yes		
Impact of IT on operational strategy			Yes	
IT influence on marketing strategy		Yes		
Use of IT systems in the company		Yes		
Objectives of IT strategy		Yes		
Thrust of IT strategy		Yes		
Management of IT in the future			Yes	
Critical success factors for competitive advantage through IT			Yes	
Management of IT projects and innovation			Yes	
Level of research and development in IT		Yes		
Nature of IT department		Yes		
Importance of IT skills in company		Yes		
Awareness of IT strategy			Yes	
Involvement of users in IT strategy			Yes	
Risks associated with IT strategy		Yes		
Periodic review of IT strategy			Yes	
Measurement of IT performance			Yes	
Characteristics of IT strategy			Yes	
Who champions IT in the company?			Yes	

outstandingly 'poor'. In general, the assessment is that the Hong Kong contractors perform 'well' in most aspects other than including IT in the formulation and delivery of the business strategy and in the marketing strategy. The use and thrust of the IT is not aligned with the goals and objectives of the company. The level of research and development is not high. IT staffing, expertise and skills are not rated highly. Possibly for that reason, the risks associated with IT are perceived as financial, or technical, and not as business or strategic risks. Evidence gained indicates that the extent of the strategic use of IT by the Hong Kong contractors is much the same as their counterparts in the UK and Australia.

Web technologies and procurement systems

Web technologies are merely tools that can, with proper training and support, provide the means for effective communications. Academics have shown by their use of the web for research through electronic libraries and the use of search engines to gather other information on the web that research activities can be effectively undertaken using the web. Additionally, the web has proved itself as a suitable medium for creating a promotional presence that can be accessed 24 hours per day, seven days a week. Persuasive arguments have been advanced that the web will be used extensively in the future by construction professionals (Walker and Betts 1997).

Construction professionals can use the web to gather estimating and forecasting information. They can also undertake other research activities on the web by using links to supplier data files on costs, to bureau of statistics data for marketing plans or to the bureau of meteorology for weather information. All such data are available outside the web but in many cases it is immediately accessible through the web. Also, the Internet can be used for promotional purposes to make potential clients aware of a company's profile, capabilities and general services offered in the same way that marketing and promotional materials currently perform this function. Again, the advantage of the web is constant access and availability – particularly if e-mail links are judiciously placed in appropriate places on a company's web site and the recipients of e-mail respond promptly and appropriately. There is something frustrating, and paradoxical, about receiving a printed brochure in the post as a reply to an e-mail generated from a web page.

Most of the companies are already using information transfer such as CAD drawings or project plans using floppy disks. Information transfer can be more quickly exchanged through the web by using file transfer technologies or as files attached to e-mails. The Internet is also being used for video conferencing at a much cheaper rate and in a more convenient setting than is currently on offer by non-net-based video conference providers.

The use of Internet or intranet access to electronic notice boards has the potential effectively to provide up-to-date project status information. There are many web sites where continuously updated photo images are being transmitted. This application of web technologies has the potential radically to change the way that information is communicated. The construction industry could make great use of this opportunity to gain a strategic competitive edge. This could be achieved in three ways. First, these technologies can reduce the cost of communication transmission through productivity gains and reduction of the need to multiple-handle information in the supply chain. Second, these technologies have the ability to allow clients already using the technologies to communicate with the construction industry by using common communication technologies. Third, if the use of these technologies is well thought through then there could be a quality of service advantage because such use offers the ability to maintain on-line current-status information of projects.

All the above advantages apply provided that use of these technologies is managed in an effective and coherent way. There are obvious dangers of poorly managed web technologies. The issue of security problems was raised earlier. Perhaps more importantly, incompetence and chaos are the greatest potential enemies of these technologies. If the client is subjected to communication breakdowns and the kind of system failures that are all too commonplace in poorly managed computer installations, then the competitive advantage becomes a competitive disadvantage. However, the technologies listed above are current technologies and really significant change will come with the adoption of visualisation technologies using object-orientated approaches which will completely change the nature of the design and construction processes. It remains to be seen how quickly the construction industry takes up these technologies, but the uptake is likely to be very fast with a consequent major change in the way procurement systems are viewed by clients. It is certain that the change will be driven by clients' demands and the imposition of clients' systems upon the construction industry. This topic was taken up by the Construct IT Centre of Excellence (http: //www.construct-it.salford.ac.uk) in its members' meeting report, 'Delivering value to clients using IT', and is reproduced below.

The contribution IT can make to improve industry performance for the benefit of clients is by:

- allowing information of relevance to client business processes to be communicated to clients as a part of the project process;
- supporting the principles of multiple parties working simultaneously on projects (concurrent engineering) through a standardised and speeded up design process allowing simulated visual evaluation of design prototypes and solutions;
- supporting the fundamental culture change required in the industry;
- allowing standard access to product data that allows dynamic access to information about product performance, simulation of component assembly and operation, and supports improved site logistics;
- supporting improved portability, accuracy and re-use of information that enables structured, remote and consistent communications, improved document and drawing management and improved flow of information;
- enabling changes to be made to supply chain relationships;
- increasing client awareness and participation in modelling of the building product, allowing costs and performance in-use

prediction, and allowing electronic capture of operation and maintenance guidelines.

The supply side of the industry must facilitate this by:

- identifying client business processes and aligning integrated project data to these;
- changing its culture to more effectively embrace change and IT adoption;
- agreeing standard formats for building product data and insisting on their use by specifiers from standardised information sources and involving a reduced number of product suppliers and following this by capturing historical data on component and product use and performance;
- agreeing standard, improved construction processes, aligning their IT systems to this and adopting and implementing standard formats for construction process information including the International Alliance for Interoperability (IAI);
- adopting a whole life-cycle view and developing IT systems that allow full life-cycle cost modelling at the design stage and supporting building and component performance assessment;
- concentrating IT application and implementation on the building product and client business processes rather than internal business processes of the supply side organisations and adopting standards including IAI;
- encouraging broad use of IT by all staff, installing standard IT systems, making greater use of the WWW [World Wide Web], and publicising their examples of IT success;
- applying manufacturing methodologies and practices from other industries by using IT to support long-term alliances, adopting electronic trading, increasing electronic interchange and involvement of the supply chain at the early stage of projects;
- integrating management tasks and reducing process waste of information management including standardising systems and processes and electronic interfaces with suppliers;
- specifying minimum IT investments and platforms for project partners.

Clients must facilitate this by:

- articulating client needs and project objectives more clearly using a more structured approach, with improved data input, and with articulated client core business processes as a part of the brief;
- insisting that manufacturers provide electronic product information and that the supply chain uses it;

- insisting on validated, maintained and transferred project product databases through encouraging openness and honesty and by insisting on and specifying electronic means of working and defined project data standards, and a policy of open electronic access to project data;
- integrating the project team into the client's organisation, by adopting supportive procurement processes with a smaller total range of organisations as project partners, involving the supply chain in internal project meetings, and electronically integrating the supply chain in client processes;
- making the flow of information within a project a priority, and facilitating it by investing in project IT and innovation;
- maintaining and communicating a focus on the full-life cycle perspective;
- defining and spreading good client experiences and practices;
- using IT tools to allow senior client decision makers to visualise project solutions at the design stage.

(Construct IT Centre of Excellence 1998)

With these changes in client orientation and expectations, coupled with the increasing power, sophistication and ease of use of IT, will come a significant restructuring of the roles of the project participants in the procurement process. Matthews (1996) gives a good account of the use of IT in promoting semi-project-partnering (see Chapter 11 for more details). Changes in design technology such as object-orientated CAD (OOCAD) will have an impact on the traditional roles of the profession and the procurement process. With knowledge-based OOCAD it is possible that one professional, be he or she the architect or quantity surveyor or project manager, will be in a position to run the whole of the design and construction process. By embedding knowledge in the objects shown in the four-dimensional CAD system the whole construction process can be analysed and value management undertaken to produce better solutions. Such advances will impact strongly on the buildability, quality and safety of projects. However, it should be borne in mind that such advances are relevant only to advanced procurement systems and will have a limited impact initially on issues such as empowerment and appropriate technologies.

Conclusions

The results of a selection of key construction industry professionals' ideas of current and near future use of web IT are encouraging from the point of view of the prospect of these technologies being offered by the industry to improve communication effectiveness. Furthermore, respondents generally support the advancement of these technologies in the construction industry as a means to provide procurement, in its wider context of project delivery, as a more competitive service to clients. Respondents expressed reasonable doubts about security, training and support issues. It was also indicated how these technologies can be used to great effect in attracting new business, in researching information for presentations and estimates and in communicating effectively between teams and the client.

At the CIB W78 symposium in Stockholm in June 1998 in his keynote speech Matthew Bacon of the British Airports Authority emphasised the need for the construction industry to take on board two key messages from the clients. These messages were:

> Standards are very important but accurate definition of client requirements is even more important. This means that if suppliers are to deliver value to their clients they must understand their clients' business processes – to help them make the right decisions – and provide them with the information that they need.
>
> Modern business is concerned with integration of business information to make informed business decisions. The Integrated Data Model (IDM) must facilitate this. This means that a wider definition of the IDM needs to be established. It must enable the client to view data from their perspective too – a perspective probably quite different from the project team.
>
> (*idem* unpublished sheet)

The pressure is on the construction industry to adapt its processes to the needs of the client; in particular, the client business and the whole project life cycle are key issues to be addressed. Hence the industry must focus not only on construction but also on facilities management. The opportunity is provided through IT to document and manage the whole life cycle, and these are key issues for the client organisation. Thus the focus of attention is on the product and the process as well as on production linked to facilities management. IT is the enabling technology which allows all of these to be brought together in an holistic approach to the procurement system.

References

Banaghan, M. (1997) 'The internet takes its place in management', *Business Review Weekly* 3 November, 88.

Betts, M. and Shafagi, M. (1997) 'A strategic IT health check', Construct IT Centre of Excellence, University of Salford, Salford; http://www.construct-it.salford.ac.uk/images/reports/Health Check/contents.htm/

Construct IT Centre of Excellence (1998) 'Delivering value to clients using IT', members' meeting report; http: //www.construct-it.salford.ac.uk

Futcher, K. and Rowlinson, S. (1998) 'Information technology used by Hong Kong contractors', *Proceedings of the CIB Working Commission W78 Information Technology in Construction Conference*, Stockholm, June 1998, 245–56.

Futcher, K. and Thorpe, T. (1998) 'The significance of data "held in context" in project information management systems', *Proceedings of the Second*

International Conference on Construction Project Management, Singapore, 19–20 February, 199–207.

Matthews, J. (1996) *A project partnering approach to the main contractor–subcontractor relationship*, unpublished PhD thesis, Loughborough University, Loughborough.

Pullar-Strecker, T. (1997) 'Introducing . . . callers who need no introduction', *Business Review Weekly* 1 December, 94–5.

Stewart, P. (1997) 'The Australian experience of the strategic IT health check', unpublished strategic health-care check report, Royal Melbourne Institute of Technology, Melbourne.

Surfas, M. and Chandler, D.M. (1996) *Running a perfect web site with Windows*, Indianapolis, IN: Que Corporation.

Walker, D.H.T. and Betts, M. (1997) 'Information technology foresight: the future application of the world wide web in construction', *CIB W78 Workshop, Information Technology Support for Construction Process Re-Engineering IT-CPR-97*, James Cook University, Cairns, Queensland.

Part IV

Procurement systems in practice

9 Applying multiple project procurement methods to a portfolio of infrastructure projects

John B. Miller

Introduction

Ostensibly new, the problems facing today's generation of governments and infrastructure planners are, in reality, quite old. These problems include:

- getting infrastructure development started or reinvigorated to improve economic efficiency and raise the standard of living;
- starting and sustaining private sector entities such as architectural and engineering consulting firms, construction companies, design and build firms, manufacturers of supplies and equipment and developers of new technology;
- building public sector institutions which facilitate economic activity, encourage competition and increase the transparency of government regulation and legislation;
- producing steady technological refreshment of infrastructure, including replacing 'dumb' with 'smart' systems, 'dirty' with 'green' systems;
- attracting both public and private sector investment of capital.

Governments continue to search for stable procurement systems which let new ideas, new technologies, new capital and new firms in, while allowing existing firms to grow and evolve. In the USA, for example, this search is as old as the nation (Miller 1996).

The rebirth of project delivery and finance as variables

The biggest news in the world of public infrastructure procurement is the rebirth of project delivery and finance as variables in infrastructure planning. Over the past decade, the engineering–procurement–construction (EPC) sector throughout the world has developed broad expertise in the full range of project delivery and finance methods, including design–bid–build (DBB), design–build (DB), design–build–operate (DBO) and build–operate–transfer (BOT). The literature confirms that, for each available delivery method, substantial knowledge, experience and judgment are required for success. After any one of these delivery methods is chosen, successful implementation of that method requires careful planning, detailed scheduling, timely materials and equipment acquisition and proper integration of all the design and construction elements.

The growing acceptance that there is more than one project delivery option for most projects is directly correlated to the growing recognition that choice of project delivery method profoundly affects the timing and scope of innovation at each step in the infrastructure delivery process. Choice of delivery method directly affects choice of technology, design approach, construction methods, facility operations and project finance. The emerging mix of project delivery and finance options necessarily implies new opportunities to package projects in order to optimise not just one but a portfolio of infrastructure facilities. This chapter explores the key elements of procurement in this new environment and offers an integrated strategy intended to align infrastructure procurement strategy with both long-term economic activity and appropriate stewardship of the environment.

Definitions

The term 'infrastructure' is used in a broad sense to mean, *collectively*:

- the transportation of people, goods and information;
- the provision of public services and utilities such as water, power and the removal, minimisation and control of waste;
- environmental restoration.

The term 'project' is used to refer generically to contracts awarded by public owners for the provision of capital works and/or infrastructure services. Using metrics I have described elsewhere (Miller 1995, 1996, 1997b) the operational framework represented by horizontal and vertical axes in Figure 9.1 was developed to describe the practical choices facing the public infrastructure sector at both the project and the portfolio levels. The horizontal axis represents the continuum of delivery methods measured by the degree to which typical elements are segmented or combined with one another; the vertical axis represents the continuum of financing methods measured by the degree to which government assumes the financial obligation for producing, operating and maintaining the project throughout its life cycle. Superimposed on the framework in Figure 9.1 are the project delivery methods in common use throughout the world. Most of these methods are described by Gordon (1994). Several variations of DBO and BOT are also included.

All the procurement methods shown in Figure 9.1 are defined in two dimensions: the means of project delivery and the means of project finance. The term 'owner' refers to the public entity procuring infrastructure facilities or services, and contractor refers to the successful bidder or proposer

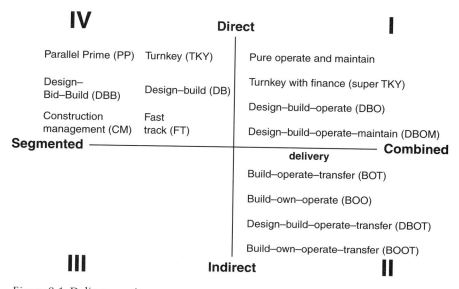

IV   **Direct**   **I**

Parallel Prime (PP)   Turnkey (TKY)   Pure operate and maintain

Design–Bid–Build (DBB)   Design–build (DB)   Turnkey with finance (super TKY)

                                     Design–build–operate (DBO)

Construction management (CM)   Fast track (FT)   Design–build–operate–maintain (DBOM)

Segmented ——————————————————— **Combined**
                                        delivery

Build–operate–transfer (BOT)

Build–own–operate (BOO)

Design–build–operate–transfer (DBOT)

Build–own–operate–transfer (BOOT)

III   **Indirect**   **II**

Figure 9.1 Delivery options
Note: horizontal axis = continuum of delivery methods; vertical axis = continuum of government finance methods
Reproduced by kind permission of the Massachusetts Institute of Technology

that emerges as the winner of the procurement process. DBB is defined to mean a segmented delivery strategy in which design is fully separated from construction. Both are in turn fully separated from maintenance and operation of the facility once the project is turned over by the contractor to the owner. In the DBB model, planning and financing of the project are also separately provided by the owner. DB is defined as a delivery strategy in which the government procures both design and construction from a single contractor. Initial planning, functional design, financing, maintenance and operation of the facility remain as separate, segmented elements of the project, provided by the owner. DBO is defined as a delivery method in which design, construction, maintenance and operation of the project are procured by the owner from a single contractor. Initial planning and functional design are provided by the owner. As in Figure 9.1, the DBO procurement method is defined to require that the public owner directly provide sufficient financing for the contractor to perform all of the tasks assigned by the owner. This financing is typically provided in one of two ways (and sometimes as a combination of the two): direct cash payments by the owner, or delivery by the owner of the equivalent of direct cash payments to the contractor, such as the right to collect user charges. BOT is defined as a delivery method in which the owner procures design, construction, financing, maintenance and operation of the facility as an integrated whole from a single contractor. Only initial planning and functional design are provided by the owner. As defined here, the BOT method puts the risk that project receipts will not be sufficient to cover project costs and debt service squarely on the contractor.

In actual procurements, owners present competing contractors with opportunities that vary slightly from these definitions. Such variations include: mixtures of direct cash payments and cash substitutes from the owner; the extent of initial planning or design; the length and extent of maintenance and operations obligations.

Public or private? The empty, useless debate

Over the past 10 years, many developed nations and international institutions that provide funding to developing nations have debated the relative merits of public and private provision of infrastructure facilities and services. Supposedly 'deep' philosophic arguments over the 'proper' role of government in infrastructure assumes that one such role can be identified – an assumption fundamentally at odds with the growing use of alternative project delivery methods. The debate is both hollow and futile. It is hollow because much of the world's infrastructure stock is already privately held, a trend that will likely accelerate, as the next major wave of infrastructure improvements – in information technology (IT) – continues to be developed primarily in the private sector. It is also hollow because much of the 'stuff' of publicly held infrastructure – aircraft, railroad cars, ships, traffic controls, highways – is not manufactured or installed by government at all but by private companies procured by government. The public–private debate is futile because public funding levels across nearly all nations seem to be inadequate to meet government's appetite for world-class infrastructure to push national economies into the global economy.

Neither a purely public nor a purely private approach to infrastructure provision has proven to be sustainable in either the developed or the developing world, particularly where financial and environmental resources are limited and where innovations in the technology and methods associated with infrastructure continue to occur unpredictably throughout the world.

The US experience is a class example of one nation's 200 years of experimenting with various mixtures of public and private procurement strategies for infrastructure. Using the framework presented in Figure 9.1, with the addition of concentric circles representing the passage of time, Figure 9.2 presents how the USA has experimented with project delivery and finance methods. Figure 9.2 summarises over 800 statutes enacted by Congress prior to 1933 which led to projects, and several thousand projects funded through federal grants after World War 2 (Miller 1995, 1996, 1997b).

The USA's current exclusive reliance on quadrant IV for federal construction of public infrastructure has, since 1980, proven to be unstable because

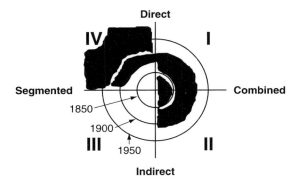

Figure 9.2 The history of US infrastructure procurement strategy
Note: for a description of the framework, see Figure 9.1
Reproduced by kind permission of the Massachusetts Institute of Technology

of chronic shortfalls in federal direct funding for infrastructure. The US experience is common to many of the world's nations, both developed and developing, with variations in timing and sequence. The 'private versus public' tug of war for the heart and soul of infrastructure development continues in the US today, as competing ideologies argue for totally public infrastructure in quadrant IV and totally private infrastructure in quadrant II.

This polarizing debate now threatens to paralyse what would otherwise be an inexorable process of infrastructure renewal – led by innovation in technology and methods and pursued by entrepreneurs and investors. The danger is that both government and private industry will continue to see procurement strategy as an ideological choice between public and private rather than as a steadily evolving mixture of both. The basic problem is how to produce and sustain a competitive infrastructure base without concentrating too much power in either the state or the private sector and at the same time maintaining incentives for individuals to innovate, to produce and to improve both themselves and the infrastructure portfolio.

The underlying logic of procurement must change

Irregular, and often unpredictable, shifts in government policy between purely public and purely private infrastructure send exactly the wrong message to designers, constructors, operators, investors, inventors and individuals interested in making their contribution to the economy through public infrastructure. Such shifts merely confirm to decision-makers in the private sector that infrastructure is not a reliable market and that more stable opportunities should be pursued elsewhere.

Procurement strategy should recognise explicitly what generations of experience has already taught: innovations enter the infrastructure portfolio through each of the individual segments in the procurement process (design, construction, finance, operations, maintenance) and through combinations of these segments (DB, DBO, BOT). Only a broad mix of procurement strategies offers a stable base for broader economic and environmental strategies. Figure 9.3 describes this strategy graphically. No single quadrant strategy (quadrant IV or quadrant II) is stable in the long term.

Fundamental elements of a sustainable procurement strategy

The fundamental elements of a procurement strategy incorporating the notion that project delivery and finance are key variables to be managed in the refreshment of the infrastructure portfolio have been previously described (Miller 1997c). These elements are listed below in summary form for discussion and debate:

- a three-quadrant strategy in which a steadily evolving mix of public and private delivery or finance is the express goal;
- consideration of alternative project delivery and finance mechanisms as one means for verifying project viability, introducing new technology and generating competition over quality, initial cost and life-cycle cost;
- comparative discounted cash flow analyses of life-cycle cost in order to permit one-to-one comparison among the DB, DBB, DBO and BOT contract awards – project scope is clearly defined in advance by the government, either through performance specifications or through design specifications;
- competition in the award of projects, through well-advertised, well-marketed requests for proposals and invitations to bidders;
- fair treatment of actual competitors, through even-handed implementation of procurement processes;

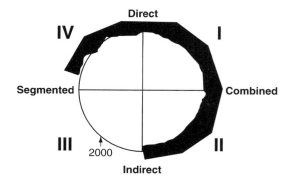

Figure 9.3 A stable procurement strategy
Reproduced by kind permission of the Massachusetts Institute of Technology

- transparency – signaling fairness to potential competitors through early statement of project and proposal requirements and evaluation criteria;
- an independent engineering check of the efficacy of design to ensure public safety whenever the design function is combined with the construction function;
- competition which is open to technological change through increased reliance on performance specifications.
- a portfolio approach to optimisation, using discounted cash-flow analyses over the project life cycle as the common denominator to compare alternative configurations of project delivery and finance for a collection of projects (Miller and Evje 1997);
- continuous integration of these concepts into government procurement strategy.

Impacts of the three-quadrant strategy on the economy, the engineering procurement construction sector and the environment

A three-quadrant procurement strategy offers numerous opportunities for continuous evaluation of new technologies, techniques and methods to improve the quality, cost and environmental performance of infrastructure services and facilities. These opportunities naturally arise once the cost of current facilities and services become transparent to operators, users, taxpayers and potential competitors, as the three-quadrant strategy allows and encourages competitive pressures to replace 'dirty' with 'clean', 'dumb' with 'smart', and to move the infrastructure portfolio steadily towards better, faster, cleaner and cheaper. The potential impacts on the economy, on the EPC sector and on the environment are significant.

The impact on innovation

The three-quadrant model permits innovations in technology to enter the infrastructure portfolio through any one of the delivery methods. The ownership, use and rewards for innovative software can be more attractive to software developers where DBO or BOT is used, encouraging more rapid deployment of IT into control systems which integrate the functions of building, plants and transportation systems. Improved energy conservation is a likely direct benefit.

With respect to innovation in architectural and engineering methods, the three-quadrant strategy permits and encourages technology transfer of methods first employed in one quadrant to subsequent projects in other quadrants. A classic example of just such a transfer is the single-gasket immersed-tube tunnel design first employed on BOT projects on the cross-harbour tunnels in Hong Kong, which has been quickly transferred to the Boston Central Artery/Tunnel project, which employed DBB as the delivery method. Innovations in construction methods are similarly transferable across the quadrants.

The impact on firms

The three-quadrant model encourages individuals to create and operate firms across the entire spectrum of activities associated with infrastructure development, including professional design, construction, DB, supply, technology, operations, maintenance, finance and combinations of these functions in broader organisations to perform DBO and BOT services. A stable, long-term commitment by government to all three of the quadrants permits and encourages new entrants and permits well-established firms to compete for new opportunities in more complex markets that integrate design, construction and technology. The three-quadrant model encourages firms to differentiate, to evolve and to focus on infrastructure as an attractive market (Porter 1980, 1985).

The impact on individuals

The three-quadrant model encourages individuals to contribute to infrastructure improvement not only in new but also in traditional ways. Numerous points of entry and advancement are maintained through continued reliance on DBB and DB as bedrock delivery methods. Associated with these methods are traditional incentives for professional design education, practice and licensure as well as solid monetary incentives for the education, training and apprenticeship of skilled craftspeople. The three-quadrant model encourages entirely new groups of individuals – technology suppliers and financiers – to participate more actively in infrastructure renewal through the integrated procurement processes of DBO and BOT. In many of these procurements, technology, software and finance will represent more important drivers in these competitions than will engineering or construction elements.

The impact on competition

The three-quadrant model encourages stronger competition among firms, not only on initial cost for DBB and DB but also on life-cycle cost, time of performance and quality of performance.

The impact on capital availability

The three-quadrant model encourages government to obtain independent checks on the economic and technical viability of large projects through DBO and BOT competitions. Such checks will gradually permit governments to adjust the allocation of projects across the quadrants so that private capital is reasonably and reliably attracted to viable projects in quadrant II, and to those projects in quadrant I where government financial backing (through cash substitutes) is clear. Private capital financing of such projects will, in turn, create new opportunities for government better to allocate direct cash payments to projects in quadrants IV and I.

The impact on portfolio planning

The three-quadrant model makes scenario analysis possible by expressly incorporating project delivery and finance alternatives into the planning process for a collection of projects, *before* procurement commences for individual projects. An early software application, called CHOICES, that applies these principles to a portfolio has been developed at the Massachusetts Institute of Technology (MIT) (Miller and Evje 1997). The purpose of the software is to permit alternative configurations of project delivery and finance to be explored by government planners before choice of delivery method is made.

The impact on government procurement institutions

The three-quadrant model necessarily implies a smaller, yet a more robust, role for government in the planning, delivery and operation of the infrastructure portfolio. Greater transparency is a substantial element of the model, including visible, reliable commitments by legislatures and regulators to a mixed delivery approach, open technologies and accurate statements of current costs and levels of performance, among other factors. In the three-quadrant model, government has a special obligation to identify the public's functional infrastructure needs, to analyse the procurement options, to select among delivery methods, to conduct the competitions and to monitor the results achieved, all as a prelude to repeating the entire process as technologies, skills and needs evolve with time. Benchmarking against similar facilities and similar services provided by other governments can help to keep this process healthy and strong. Steady, sustainable improvement in the portfolio is the strategic goal.

Portfolio planning and procurement with project delivery and project finance as variables

The existence of delivery alternatives across a group of desirable projects and public sector capital constraints creates the opportunity to optimise the placement of individual projects in the quadrants shown in Figure 9.1. The portfolio planning problem, represented schematically by Figure 9.4, is how systematically to analyse and choose from a number of configurations of the portfolio, all of which meet capital constraints in each year of the duration of the analysis.

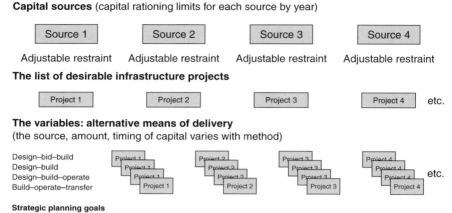

Figure 9.4 Choices: portfolio configuration options
Reproduced by kind permission of the Massachusetts Institute of Technology

Discounted cash flows for each distinct portion of the project's life cycle provide the common analytical tool to compare and contrast different configurations of the portfolio. The ability to compare and contrast alternative configurations will, in turn, offer attractive new opportunities for scenario analyses in advance of procurement decisions. Alternative configurations of projects permit decision-makers to examine and predict how changes in the amount, timing and source of funds affect the entire portfolio.

Project discounted cash flows

Discounted cash-flow (DCF) models have been used for years to evaluate the return on single and multiple alternative investments. Initial cash flow is negative at the time of the investment, which is typically assumed to be made in the first time period of analysis. After this initial investment the revenues which follow are used, after appropriate discounting, to determine whether the decision to invest is sound in the first place. The magnitude of the revenues in each subsequent time period, together with the selection of an appropriate discount rate, establishes the predicted net present value of the investment (Brealey and Myers 1996).

In the construction industry, contracts for design, construction, operations and finance (and combinations thereof) supply a snapshot of the financial relationships between the parties. These contracts capture the parties' mutual agreement as to cash flow. For DBB contracting systems, Figure

9.5(a) provides a typical picture of the resulting cash flow, with an interval of reduced negative cash flow between the design and construction phases, as the project is bid and a general contractor selected. Figure 9.5(b) shows a slightly different profile for DB, where there is no interval between design and construction and a slightly smaller cash stream is expended more quickly and over a shorter construction duration than in DBB.

Figure 9.5(c) presents a typical discounted cash-flow scenario for DBO. In the example presented, the owner has committed to a stream of payments during operation that is sufficient, if design and construction are on schedule and within budget, to produce a positive cumulative DCF over the contract period. Figure 9.5(d) presents a typical DCF scenario for a pure BOT scenario, where, after initial planning and award by the owner, all financial obligations and rewards are assumed by the franchisee throughout the franchise period. Figure 9.5(e) shows a typical DCF scenario for a pure operate and maintain (O&M) agreement, typical of current arrangements controlling O&M services on many waste water treatment plants in the USA.

The arrows to the right of the cash-flow stream, marked by question marks, indicate that at the end of the contract period, neither the public owner nor the private contracting party is bound to continue with the pattern of DCF allocations mutually adopted in the original agreement. The arrows indicate that a decision point has been reached (typically at the end of a contract obligation). The government must now select a new method of acquisition, based upon yet another snapshot of cash flow embodied in a new project delivery method and a new contract obligation.

Project delivery and finance are variables which significantly affect the time, cost and quality of infrastructure facilities, yet evaluation models typically do not consider these effects (Brealey and Myers, 1996; Niemeier *et al.* 1996; Tigue 1996). Using project-specific models for real collections of projects, explicit comparison of alternative combinations of the life-cycle discounted cash flow associated with each of the five basic delivery methods can be made. Since DCF analyses are linear models of financial performance, the summation of cash flows across any given period provides linear constraints for the portfolio, and the summation of net present value for all projects equals the net present value of the portfolio (Brealey and Myers 1996).

Through combinations of the DCF profiles set forth above, most existing infrastructure facilities can be effectively modelled. For example, the Wilmington Wastewater Treatment Plant in Wilmington, DE, began its life cycle as the product of 90 per cent federal and state subsidy pursuant to the US Environmental Protection Agency (EPA) Construction Grants Program (EPA 1972, 1975). Figure 9.5(a) represents a typical cash flow for this type of US federally reimbursed DBB construction project. After years of public operation with public employees provided from sewer fee collections, the next stage in Wilmington's life was a pure O&M contract, schematically represented by Figure 9.5(e). In 1996 the City experimented with converting

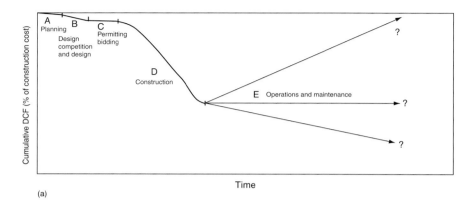

(a)

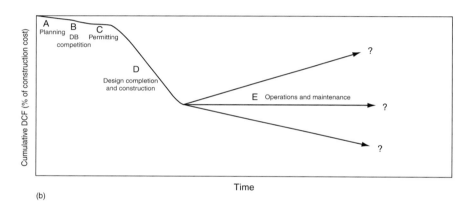

(b)

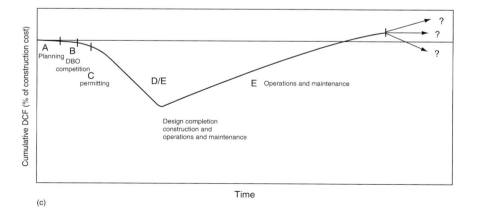

(c)

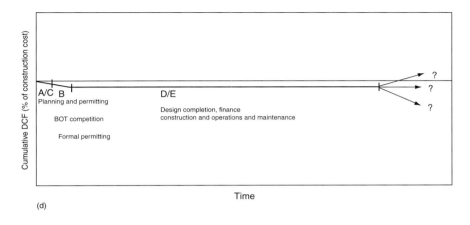

(d)

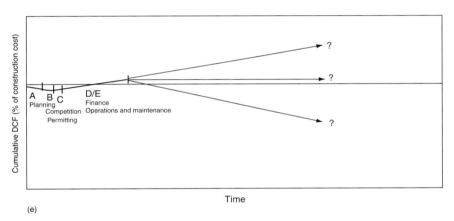

(e)

Figure 9.5 Typical cash-flow profiles: (a) design–bid–build (DBB); (b) design–build
(DB); (c) design–build–operate (DBO); (d) build–operate–transfer
(BOT); (e) operate and maintain
Note: DCF = discounted cash-flow; question marks represent decision points
Reproduced by kind permission of the Massachusetts Institute of Technology

the project to BOT, a scenario which would follow Figure 9.5(d). The combination of these procurement delivery strategies is represented in Figure
9.6 (Miller 1997a).

An outline of 'choices'

This section describes the software developed at MIT which attempts to use
DCF models of life-cycle costs as an effective decision tool to manage infra-
structure portfolios. The CHOICES software is the property of the
Massachusetts Institute of Technology, © 1997 and 1998.

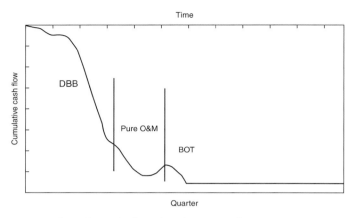

Figure 9.6 Life-cycle cash flow for the sum of project delivery choices: design–
bid–build (DBB) plus pure operate and maintain (O&M) plus build–
operate–transfer (BOT)
Reproduced by kind permission of the Massachusetts Institute of Technology

Portfolio assessment

Projects are assessed, planned and procured in the context of a portfolio of
infrastructure projects. Figure 9.7 shows the concluding summary of historical receipts and expenses of a New England town, including not only
the procurement and maintenance of the infrastructure portfolio but also
all the sources and uses of funds by the town for a significant period in the

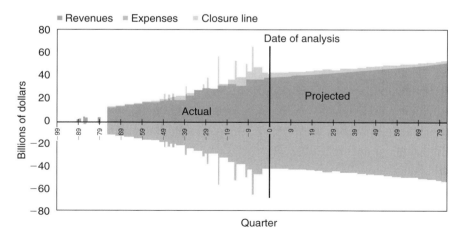

Figure 9.7 Portfolio (revenue and expense) assessment for a New England town: the
context for project delivery choices
Reproduced by kind permission of the Massachusetts Institute of Technology

past (in this case over 15 years). This information is presented to the left of the line labelled 'Date of analysis' in the figure. Figure 9.7 represents the sum of several spreadsheets which separately track all the major components of the town's receipts and expenses by function. Historical receipts and expenditures by function are used to project future average receipts and expenditures to the right of the 'Date of analysis' line (each of which can be modified by the analyst to adapt the data to a particular infrastructure portfolio).

This context provides construction professionals with a clear understanding of the relative significance of the infrastructure portfolio in the overall economic performance of the government, the historical expenditures of government on each major function, including infrastructure, and the historical pattern of government spending on capital debt, maintenance and operations.

Project modelling and scenario analyses

Superimposed on this historical context is the task of analysing how each of the various projects desired by government might be delivered, using project cash flow and project delivery method as variables. The approach followed in CHOICES is to develop multiple delivery alternatives for each project in the portfolio and then to explore a variety of possible configurations of the portfolio based upon the summation of resulting cash-flow and resource constraints.

The problem is represented schematically by Figure 9.8. Project delivery options A through E are considered for application to *n* projects. A single time period (typically a quarter) is shown in Figure 9.8. The first step in the CHOICES model is to develop pro forma DCF models for each project in the portfolio and for each of the delivery methods which appear to be feasible. In many cases, it is readily apparent that one or more of the five

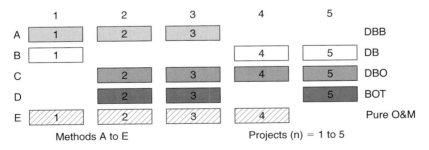

Figure 9.8 Viable methods kept
Note: DBB = design–bid–build; DB = design–build; DBO = design–build–operate; BOT = build–operate–transfer; O&M = operate and maintain
Reproduced by kind permission of the Massachusetts Institute of Technology

delivery methods shown in Figure 9.8 is not practical. For example, BOT methods are simply unworkable for local street repair and maintenance, unless government is willing to permit BOT franchisees to collect fees from users. DBB may be the only acceptable delivery method to government for a project in which architectural considerations are paramount, for example, the renovation of Westminster Abbey. DBO may be deemed to be impracticable for a prison, if government concludes that the private sector should not be in the business of incarceration. BOT may prove to be impractical for a toll road, where competing free roads put commercial viability of the toll road in grave doubt.

The development of cash-flow models quickly demonstrate that one or more of the project delivery alternatives is not practical, because of inadequate revenue sources, a mismatch in project size or complexity or the availability of appropriate technology. The generation of a solid financial pro forma requires a working knowledge of the project, how it works, how it will be funded and how much it is expected to cost, both for construction and for operations. The alternatives that are not viable are discarded from further consideration, as shown in Figure 9.8, using the general approach set forth by Gordon (1994), with extensions of the theory for application to BOT and DBO.

The next step in the CHOICES analysis is the expansion of the two-dimensional model shown in Figure 9.8 to three-dimensions, as shown in Figure 9.9. Time is added along the third axis to permit annual capital budget constraints to be checked and applied across trial configurations of the project portfolio. In each time interval, the summation of discounted cash flows must not exceed capital constraints during the period. Each project delivery option has already been independently checked for commercial and

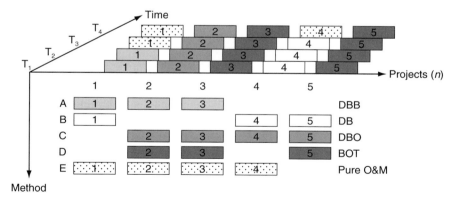

Figure 9.9 The chooser function in CHOICES (© 1997 and 1998, Massachusetts Institute of Technology)
Note: for abbreviations, see Figure 9.8
Reproduced by kind permission of the Massachusetts Institute of Technology

technical viability, leaving only those viable methods left for optimisation. Called the 'chooser function' in the CHOICES model, the section of the model collects quarterly cash-flow information for each viable delivery alternative and assists decision-makers in the process of sifting through the numerous alternatives.

The CHOICES model includes:

- the DCF life-cycle cost models for each project (and all the alternative delivery strategies);
- summary sheets containing the DCF streams for those project-specific delivery alternatives deemed to be viable and worthy of further analysis;
- the chooser function, which uses mathematical programming techniques to sort through the possible combinations in a structured way;
- presentation graphics, which report the identity, project delivery mix, cumulative discounted cash flows associated with various configurations of the portfolio, and shortfalls derived from the capital budgeting constraints.

Figure 9.10 shows the results in the chooser function of just one configuration of a collection of projects for a small New England town. Actual receipts and expenditures are shown to the left of the white vertical line, and projected town expenditures for all functions (including the portfolio of infrastructure projects) are shown to the right of this line. The cumulative project cash flows for this particular configuration of projects are added to the projected receipts and expenses derived during the portfolio

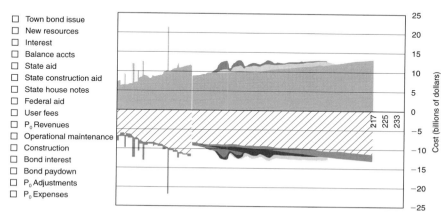

Figure 9.10 Portfolio management: one scenario for a New England town – a comparison of revenue stream with costs, by quarter

Note: P_0 = project 0; it refers to the results illustrated in Figure 9.7 and are revenues received and expenses made outside the portfolio of projects being analysed in CHOICES

Reproduced by kind permission of the Massachusetts Institute of Technology

assessment stage. In Figure 9.10, receipts and expenses are reduced to the right of the white vertical line because, in this hypothetical scenario, the town no longer provides water and wastewater services. These receipts and expenses (though reduced in absolute amount) are transferred directly to users of these services. Figure 9.10 shows only just one of many possible configurations.

Conclusions

The growing acceptance of multiple project delivery and finance methods necessarily implies that governments will be increasingly faced with strategic choices whether to use 'public' or 'private' mechanisms in the provision of infrastructure facilities and services. History teaches that purely public and purely private delivery mechanisms are unreliable, unstable and averse to innovation. Steady, sustainable improvement in the infrastructure portfolio will be achieved through a transparent, mixed strategy – a three-quadrant model – which encourages individuals and firms to innovate, which encourages technology developers and investors to enter and which is simple for participants to understand and use. A flexible, reliable, mixed public–private procurement strategy is required if broader questions related to the economy and the environment are to be coherently addressed through procurement systems. Governments, at all levels, are now in a position to use alternative delivery mechanisms to make dramatic improvements

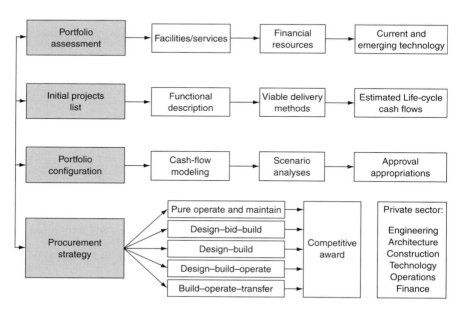

Figure 9.11 The emerging public–private infrastructure strategy
Reproduced by kind permission of the Massachusetts Institute of Technology

in the technology used in the infrastructure portfolio, in the quality of infrastructure services and in cash flow.

Figure 9.11 shows how this planning process will evolve over the next few years. Government's ability to use the procurement process to deliver efficiently and effectively infrastructure services and facilities with higher cost and quality performance is near at hand. Project delivery method and project finance method are the tools government can use to drive these improvements continuously.

References

Brealey, R.A. and Myers, S.C. (1996) *Principles of corporate finance*, New York: McGraw-Hill.

EPA (1972) *FWPCA conventional treatment rules*, volume 40, Code of Federal Regulations, part 35, section 970, appendix E, pages 4–6, Washington, DC: US Environmental Protection Agency.

EPA (1975) *FWPCA rules*, volume 40, Code of Federal Regulations, parts 30, 33, and 35; also in volume 40, Federal Register, page 20232; Washington, DC: US Environmental Protection Agency.

Gordon, C.M. (1994) 'Choosing appropriate construction contracting method', *Construction Engineering and Management* **120**(1): 196–210.

Miller, J.B. (1995) *Aligning infrastructure development strategy to meet current public needs*, PhD thesis, Massachusetts Institute of Technology, Cambridge, MA.

Miller, J.B. (1996) 'Toward a new American infrastructure development policy for the 21st century', *Infrastructure* **1**(3): 1.

Miller, J.B. (1997a) 'Case study IDS-97-W-104: DBO competition for the Wilmington (Delaware, US)', *Waste water treatment works*, Cambridge, MA: Massachusetts Institute of Technology.

Miller, J.B. (1997b) 'Engineering systems integration for civil infrasturcture projects', *Journal of Management in Engineering, ASCE*, (September/October): 5.

Miller, J.B. (1997c) 'The fundamental elements of sustainable procurement strategies for public infrastructure', *First International Conference on Construction Industry Development*, Singapore, 383–90.

Miller, J.B. and Evje, R.H. (1997) 'Life cycle discounted project cash flows: the common denominator in procurement strategy', *First International Conference on Construction Industry Development*, Singapore, 364–71.

Niemeier, D.A., Tracy, L., Reed, G., Rutherford, S. and Morin, P. (1996) 'Statewide programming: implementing transportation-policy objectives', *Journal of Infrastructure Systems* **2**(1): 30–9.

Porter, M.E. (1980) *Competitive strategy: techniques for analyzing industries and competitors*, New York: The Free Press.

Porter, M.E. (1985) *Competitive advantage: creating and sustaining superior performance*, New York: The Free Press.

Tigue, P. (1996) *Capital improvement programming: a guide for smaller governments*, Chicago, IL: Government Finance Officers Association.

10 Multiple performance criteria for evaluating construction contractors

*Mohan M. Kumaraswamy and
Derek H.T. Walker*

Introduction

The scope of this chapter has necessarily been framed within the aims and objectives of this publication and shaped by the required complementarity with other chapters. It begins with an assessment of the need for developing and clearly conveying multiple selection criteria for choosing project participants. While the selection of design and construct contractors and consultants will be briefly compared, the focus will be on the selection of construction contractors, given the more serious concerns in this area. It then reviews a sample of suggested multiple selection criteria in the context of the identified needs for more comprehensive, rigorous and transparent tender selection processes. An example is given of a hypothetical project for a simple demonstration. This is followed by a discussion of research findings on the evaluation of specific criteria such as construction time performance and the potential claims and disputes that could in turn affect cost performance. This focus on critical cost and time criteria leads to the importance of assessing the past and potential management of project team relationships during both the design and the construction processes.

This chapter thus aims to draw attention to a range of methodologies for incorporating broader contractor selection criteria, so as to enhance the targeting of overall project success. Although this chapter has been written primarily in the context of contractor selection, the methodologies discussed can apply equally to the selection of subcontractors and consultants. The envisaged practical outcome for construction clients, practitioners and students of construction management is to provide guidance on the assembly and use of such multiple selection criteria for assessing potential performance of a contractor and also to boost the probability of overall client satisfaction with the construction project and the resulting product.

Multiple needs of construction clients

Despite the customary cautionary clause that the lowest tenderer need not be awarded a contract, it is usual for the contractor to be chosen on the basis of the price criterion alone. However, it is not unusual for clients to regret such decisions in hindsight, for example when besieged by claims for extra costs or beset by poor performance levels. Enlightened clients have adopted measures such as contractor registration (where applicants are assessed in advance by clients with large and continuing project portfolios) and prequalification (for large and/or special projects) in order to screen and shortlist tenderers for a given project, prior to pricing of tenders. However, it has been proposed that the consideration of non-price criteria should extend further to the final evaluation of tenderers (for example, see Holt *et al.* 1993; Kumaraswamy 1996; Hatush and Skitmore 1997a).

Suitable selection methodologies for choosing construction participants themselves constitute an important subsystem of appropriate procurement systems that could contribute to desired project performance levels (Kumaraswamy and Dissanayaka 1996). Such selection should thus be based on a demonstration of potential to perform better against broader definitions of project success than just capital cost. For example, life-cycle costs and quality or fitness for purpose must also be considered when evaluating designers' work, as well as that of contractors who submit design alternatives or those who tender on a design and constuct basis.

Exercises to identify and express the client's expectations to potential participants are thus necessary. While the precise needs and their relative importance will vary from project to project, certain common criteria may be established for certain types of projects and participants such as for contractors in 'traditionally' procured building projects. Multiple contractor selection criteria have been previously proposed in more detail for prequalification of tenderers (for example, see CIDA 1993a; Russel *et al.* 1992). These should broaden the scope of performance potential assessments, for example to include functional and construction time performance as well as economic performance. An organisation's business performance would affect its stability and hence its capacity to complete a project, as would its experience in teamworking in multiparticipant projects and environmental and safety matters. Broad performance measures (indicators) and historical data banks thus need to be developed to benchmark potential participants.

Clients also need to demonstrate the transparency of the selection process so that the successful tenderer is clearly seen to have presented a superior submission. This is needed for clients to be accountable to project stakeholders and the broader community. It is also needed for development of the industry: since a clear indication of the goalposts would enable tenderers to target them. Feedback would facilitate continuous improvement towards current client concerns. A recent survey by Holt *et al.* (1996: 109) confirmed that 'any system that promoted constructive feedback to unsuccessful tenderers would encourage firms to address their identified areas of weakness'. Feedback to clients would in turn facilitate organisational learning that could direct future procurement strategies towards enhanced performance. The importance of establishing clear criteria and performance measures thus extends beyond improved project monitoring and control of performance levels, to learning lessons for future improvements (Kumaraswamy 1991; Kumaraswamy and Thorpe 1996).

Multiple selection criteria

Translating needs and critical success factors to measurable criteria

The translation of multiple needs of construction clients to corresponding selection criteria appears to have been inadequate. It is necessary for clients and their advisers to formulate a framework of explicit criteria that expresses both the general and the particular needs of the client and the project in the context of the given scenario, including market and environmental conditions as well as industry and community concerns.

A recent exercise for indentifying 'universal criteria for prequalification and bid evaluation' by Hatush and Skitmore (1997a: 36) indicated common criteria of financial soundness, technical ability, management capability and the health and safety performance of contractors. Four main criteria of finance, personnel, technology and (organisational) experience have also been proposed (Kumaraswamy 1997a), with sets of subcriteria and corresponding indicators which could be used for measuring against such subcriteria. Further criteria were then suggested, depending on the targeted distribution of the full portfolio of project risks, namely, design, construction, managerial, financial, market and general.

In such a context of considering overall project success, a recent study of design and build project 'success factors' by Songer *et al.* (1996) spanned a sample of 137 owners in the USA and UK. Table 10.1 indicates the derived rankings, ranging from the most important (1) to the least important (6).

Hatush and Skitmore (1997b) also investigated the perceived relationship between 20 contractor selection criteria considered at prequalification stage and project success factors in terms of time, cost and quality performance levels. They identified the dominant contractor selection criteria as: past failures, financial status, financial stability, credit ratings, experience, ability, management personnel factors and management knowledge.

The next step would be to identify indicators (measures) for evaluating companies against such criteria. For example, in relation to the first few dominant contractor selection criteria cited immediately above, a literature search yielded a proposed methodology for predicting construction company failures (Abidali and Harris 1995) as well as financial ratio models for improved contractor evaluation (Edum-Fotwe *et al.* 1995).

Table 10.1 Example of ranking of success criteria on design and build projects

Success criteria	Combined ranking	US ranking	UK ranking
On or under budget	1	1	1
On or before contracted completion date	2	3	2
Meets or exceeds the user's envisioned functional goals (fitness for purpose)	3	2	3
Meets or exceeds all technical performance specifications provided by the owner	4	4	4
Meets or exceeds the accepted standards of workmanship in all areas	5	5	5
Construction process does not unduly burden the owner's project management staff	6	6	6

Note: 1 = most important, 6 = least important

Re-evaluating selected criteria used at registration and prequalification stages

From the examples in the previous two paragraphs, it also appears beneficial to delve deeper into the more developed systems (and the more detailed proposed systems) for evaluating the technical capabilities of contractors at the registration and prequalification stages. It may then be found advantageous to incorporate or reconsider these criteria at the final tender evaluation stage as well, as indicated by Holt *et al.* (1993). This is contrary to some previously held views in the industry, that prequalification provides a list of capable tenderers who should thereafter be judged merely against price criteria. It should now be considered that prequalification is just a minimal first step, that is, a necessary but insufficient condition to establish organisational credentials. Post-qualification is also needed, since the circumstances of prequalified organisations can change considerably in terms of other work in hand, senior staff availability or financial capacities. For example, such changes led to the rejection of the prequalified lowest tenderer on a prestigious bridge project in Hong Kong.

Most large governmental or government-sponsored clients evaluate and register contractors in graded categories according to their evaluated capacities to execute contracts up to a certain value, such as: up to x million, between x and y million [which may in turn be split into many more bands (as by the Insitute for Construction Training and Development, which maintains the centralised registration lists in Sri Lanka], above y million, etc.

For example the Hong Kong Housing Authority maintains its own list, re-evaluating registered contractors regularly. It is also upgrading its evaluation criteria, for example to incorporate appropriate weightings for contractors' various resource levels and organisational arrangements, in addition to standard registration criteria based on financial capacities, organisational track record or experience and factors relating to key personnel.

Prequalification is carried out for a special project and may thus include both general and project-specific criteria. Of course, general guidelines for particular countries or work categories may be developed from best practice, for example by the Australian Construction Industry Development Agency (CIDA 1993a). Comparisons of common prequalification criteria between Saudi Arabia and the USA have indicated similar rankings of perceived importance (Bubshait and Al-Gobali 1996).

Decision aids to assist clients in prequalification processes have also been developed, for example a knowledge-based (expert) system by Russel and Skibniewski (1990) in the USA. A case-based reasoning framework was proposed by Ng *et al.* (1995) based on prequalification practices in the UK. Russel and Skibniewski's system was based on evaluation criteria such as candidate contractor's reputation, past performance, financial stability, experience record, firm capacity, current workload and technical expertise. A consideration of the possible status changes in many of the foregoing factors between prequalification and contract award in the particularly volatile and 'mobile' construction industry underscores the need to re-evaluate the contractor's status at the tender evaluation stage.

Some dangers in selecting the lowest tenderer

The need for re-evaluation (at tender stage) of factors identified in the foregoing subsection appears increasingly important, especially as it is the most recent performance, workloads and capacity status that could bear on the new project. Lowest tender cost has been identified as a crude and unsatisfactory measure when used as a sole criterion for awarding contracts and for measuring project success (Latham 1994). A construction contract may be won on the basis of low tender cost owing to incorrect cost estimation or poor risk assessment. Alternatively, the contractor may have astutely identified significant design inadequacies, potential changes in scope and contract conditions that would pose a high risk to the client in terms of receiving contract claims (Yogeswaran and Kumaraswamy 1997). The contractor can then attempt to generate high profits having won the project on a low tendered profit margin through these claims (Latham 1994; Nahapiet and Nahapiet 1985). Even if the contractor is not seeking to maximise profit through generating contract claims, a misguided client focus on suppressing initial costs may well lead to subsequent variations caused by substantial changes in scope or design modifications and/ or refinements. Contract claims have been shown to be a major cause of disputes on projects that also generate an environment of escalating mistrust, poor team relationships and client dissatisfaction (Kumaraswamy 1997b). Additionally, the least tendered capital cost may not be the most economical option in the longer term. Tenderers now frequently offer design alternatives and/or lower construction time frames, both of which also make the process of tender evaluation more complex. Thus construction cost as a sole criterion is a particularly unsophisticated means of measuring the increasingly sophisticated options available to clients.

Tendered construction time frequently influences a tender result. In demonstrating potential performance a tenderer can submit a project plan and also provide other benchmark measures of construction time performance (CTP). Results of a series of recent studies undertaken in Australia in which 64 projects were investigated (Walker 1997b) provide a useful methodology not only for benchmarking CTP but also for helping to understand the reasons why some buildings are constructed more quickly than others. Research in Hong Kong yielded useful information on the relative significance of factors that affect construction project duration (Chan and Kumaraswamy 1995). The literature on CTP includes reports on a number of studies undertaken in various parts of the world, for example by Bromilow *et al.* (1980) and Ireland (1983) in Australia, Nahapiet and Nahapiet (1985) in the USA and Kaka and Price (1991) in the UK. Coarse measures of CTP have been derived, with time in workdays being related to scope in terms of cost and/ or gross floor area in such studies, with pointers to other factors that could fine-tune predicted durations.

Fitness for use or fitness for purpose is another important measure of performance. This measure goes beyond the traditional concept of quality, in that a project can be designed to very high-quality levels but may still be of low utility or functional value to the user. An example of this would be a classroom space that is superbly designed for ease of use, with state-of-the-art communication multimedia equipment. If, however, the available expertise for servicing and supporting the facility is woefully inadequate, then the equipment may soon be unusable. Many such examples of shortfalls in performance expectations can be cited. Thus planned quality performance indicators can be misleading without a building end-user evaluation that measures the project's fitness for purpose. This particular criterion would be specially important in selecting design and construct tenderers, or where design alternatives are offered.

Potential quality management expertise is another important criterion that can be demonstrated by tenderers in many ways. They could rely on national quality assurance standards (for example British or Australian) or international (ISO 9000) accreditation of their quality practice and procedures. They could also arrange for inspection of their past and present work to demonstrate attainable standards that can be expected of their work team. Other performance measures may include evidence of winning regional or national excellence awards.

The foregoing discussion has centred on limitations in performance measures for cost, time and quality which have been traditional indicators of client satisfaction. However, other performance measures have emerged as useful. For example, the management experience of the contractor with similar types of project, teamworking relationships (both at an operational and at a strategic level), geographical location of the contractor in relation to the project site and the pricing of basic rates are all commonly used performance measures. However, the tenderers' ability to make a reasonable profit that ensures business continuity is a measure that only the commercial entity proposing to undertake the work can assess (unless the payment method is cost plus), since the true costs and profit margins are commercially sensitive. The client, however, should also have an interest in the commercial viability and expertise of those it contracts with, to realise a project (Holt *et al.* 1994).

Apart from satisfying such direct and indirect needs of the client and the contractors there should be consideration for the needs of the general community. Performance measures should be designed to evaluate transparency, accountability, sound corporate governance, ethics and social and environmental concerns. They should also target long-term technological transfers or exchanges, personnel training and industry development (Kumaraswamy 1997c).

Social and environmental concerns have also surfaced in demands to incorporate measures of health and safety performance levels on previous projects, for example through a contractors' occupational health and safety (OHS) record. Safety, health and environment (SHE) plans for the proposed construction project (European Construction Institute 1992) can also be required of tenderers, along with method statements. All of these can be verified during project delivery. Evidence of a waste-minimisation programme is just one such example of a proposed environmental management plan that can be required during procurement (Graham and Smithers 1996).

Some innovative approaches to contractor selection

A scan of recent approaches that recognise at least some of the emerging evaluation criteria and that minimise the dangers highlighted in the foregoing two subsections reveals:

- the rationalisation of contractor registration criteria and indicators (measures) that facilitate domestic contractor development and longer-term industry benefits in less developed construction industries (Kumaraswamy 1997c);
- an insistence on ISO 9000 certification as one of the prerequisites for registration in governmental lists in Hong Kong (introduced in phases) coupled to rules and guidelines for administering lists of approved con-

tractors, to be used for reviewing their performance, including specific performance requirements on safety aspects and for regulatory actions for poor performance, as indicated in the relevant Hong Kong Government technical circulars issued by the Works Branch, now renamed the Works Bureau (for example, Works Branch 1997);

- special initiatives such as the performance assessment and scoring system (PASS) for contractors registered with the Hong Kong Housing Authority; this provides for more tendering opportunities to listed contractors who perform better on recent or ongoing projects, as evaluated under the comprehensive and periodic PASS reviews (Kumaraswamy 1996);

- the contrast between the foregoing mechanism (in Hong Kong), to reward better performance with increased tendering opportunities, and an alternative approach to provide a tendering price advantage (up to a maximum of 5 per cent) for contractors who achieved higher construction quality assessment system (CONQUAS) scores in recent contracts in Singapore (CIDB 1995);

- the further contrast between the above approach and 'alternative contracting' experiments in the US Army Corps of Engineers and end-result-specification-based payments in the US New Jersey Department of Transportation, where contractors may be paid up to 103 per cent of the contract value for exceeding the 'statistical' (statistically expressed) specification (Tarricone 1993);

- 'A + B building methods' in other highway projects in the USA, where tenderers bid for both the cost part (A) and the time part (B); the lowest combination of A + B winning the contract; Herbsman (1995) demonstrated how the time bid (in days) is translated into dollars using road-user cost factors (expressed in dollars per day) and then analysed 101 contracts that were awarded using this method and found substantial time savings, with almost no additional cost, in comparison with similar projects using a cost-only criterion [of course, appropriate client-specific weightings may be incorporated (in advance) before combining cost and time criteria on different types of projects];

- an average-bid method where contracts have been awarded to the contractor whose tender price is closest to the average of all submitted bids, as occurs, for example, in Italy and Taiwan (Ioannou 1993), countering the dangers of awarding a contract to a tenderer who either mistakenly or deliberately submits an unrealistically low bid; variations in this approach are possible, for example, in selecting either the closest to the arithmetic mean or the closest below the arithmetic mean and other systems may also reject tenders that fall outside a certain band (say, 15 per cent on either side of a client's (or engineer's or quantity surveyor's) estimate;

- initiatives to identify factors affecting contractors' performance in different scenarios (Assaf *et al.* 1996; Tam and Harris 1996) and to incorporate these into models for predicting performance; these could in turn be used for refining criteria and indicators needed for evaluating potential contractors at the tender stage.

Some lessons learnt from evaluating consultants

Along with the now perceptible but gradual shift away from purely price criteria to incorporate a range of other considerations in evaluating contractors, as reflected in the foregoing subsections, there has been an opposite shift from purely technical (reputation-based or capacity-based) criteria to include price-based (fee-related) comparisons when selecting consultants for design and supervision. While the economic and social needs for the latter type of comparisons have also been acknowledged, for example by the Construction Industry Council in the UK, it is warned that while professional fee scales are no longer relevant or appropriate, cheapness is seldom synonymous with value for money (CIC 1992). Similarly, the US Supreme Court ruled against restrictions on professional price competition as far back as 1978, although the relative ethics of different degrees of reasonable price competition among consultants are still being debated (Stout 1995).

In other examples of conflicting concerns, the chairman of a leading international engineering consulting organisation urged a limit to reduced fees and corrspondingly lower staffing levels by 'learning to say no' to clients in order to arrest or reverse dropping standards (Osborne 1996). An ongoing CSIRO survey in Australia is investigating the causes of reduced quality of building design and documentation and its impact on the effeciency of construction projects following perceived links to reduced fees (Civil Engineers Australia 1997).

Such concerns and criteria – as developed for evaluating the capacities and competencies of designers and consultants – are useful in developing the criteria and indicators to assess the technical capabilities of contractors. This is even more relevant in design and construct scenarios or where alternative designs and/or construction methods are submitted for evaluation. Specific approaches have been described, for example, by Potter and Sanvido (1995) in design and build prequalification guidelines and by Yng *et al.* (1997) in a model for the selection of consultants by design and build contractors.

Criteria, such as life-cycle costs of facilities to be constructed, may naturally assume more importance in design and build construction scenarios, along with functional, aesthetic and environmental concerns. The convergence of technical and economic concerns and criteria used in different forms of procurement thus appears to provide lessons that could enrich evaluation processes for assessing various potential participants. Such insights are thus useful in developing a pilot model for contractor selection using multiple criteria, as discussed in the next section.

A worked example to illustrate the basic use of multiple selection criteria in evaluating tenders

Basic methodology

The range of potential performance criteria and indicators surveyed in the previous section indicates the complexity that would discourage the design of a universal evaluation system. However, a basic model and guidelines may be demonstrated so that project-specific criteria and indicators may be developed in each case. The starting point would be first to define stakeholders and their project objectives. This leads to the establishment of project goals. A procurement committee of project stakeholders can next develop appropriate criteria and indicators (measures) related to these goals. These performance measures can then be evaluated to arrive at indicators of project success from multiple points of view, similar to other forms of evaluation matrices proposed by Holt *et al.* (1994) and by Griffiths and Headley (1997).

Multiple selection criteria analysis can be undertaken, based upon a rating (*R*) for each candidate against each performance criterion in turn. These ratings would be weighted by the perceived importance (*I*) of each criterion in terms of impact upon the client, the end user or both. The product of *R* and *I* thus provides weighted ratings (scores) which can be used to measure the expected relative performance levels of the different candidates against each criterion. Summation of the weighted ratings of each candidate for all criteria would indicate the overall relative strength of each candidate.

A single indicator of potential performance can thus be established to form the basis for transparency in a tender selection process. Additionally, results from this analysis can be intelligently used as the basis of both sensitivity analysis and futher negotiations, if permissible and useful in the given context.

Worked example

The following simple scenario is developed as a basic demonstration example. A client has shortlisted five tenderers for a project. Table 10.2 presents the performance measures against selected criteria for assessing the tenderers. Factors rated on a basis of 1 to 5 can be converted to an index number where a 5 is represented by the index 5/5 (1.00). A tenderer would then need to submit a wide range of information including demonstrated evidence of performance capabilities not usually supplied at tender stage. (Even if supplied earlier, for example at registration or prequalification stages, such information is not usually updated at the tender stage).

A zero score would be possible against certain criteria, for instance if the tenderer made no attempt to demonstrate performance in any of the

Table 10.2 Examples of performance criteria and measures

No.	Criterion	Measure
1	Capital cost (tender sum)	Lowest tendered sum/or (this) tender sum
2	Annual operating cost (for design and construct projects or for alternative designs)	Lowest annual operating cost or (this) tenderer's proposed annual operating cost
3	Construction period	Lowest construction period or (this) tenderer's proposed construction period
4	Construction quality potential	Perceived rating
5	Concern for safety, health and environment (SHE)	Perceived rating
6	Management expertise	Perceived rating
7	Management experience with similar type projects	Perceived rating
8	Teamworking relationships	Perceived rating
9	Community support experience and corporate ethics	Perceived rating
10	Claims and disputes avoidance potential	Perceived rating

Note: perceived ratings vary from 1 (very low) to 5 (very high)

criteria numbered 4 to 9 in Table 10.2. The evaluation against criterion 10 may be subjective, based on past organisational reputations and evaluation panel perceptions of the personnel to be deployed on the new project. However, this evaluation may also be more structured, using methodology to be described in a subsequent section.

The client's evaluation panel would predetermine the criteria and also the importance weightings for each of the identified criteria in the particular project, based on the client's priorities in meeting identified stakeholder needs. For example, if the project requires community consultation and/or a higher degree of involvement or interaction, then criterion 9 may be considered worthy of a higher importance level. If costs must be kept strictly within budgeted levels and/or disputes must be contained, criterion 10 may be assigned high importance as well.

The determination of criteria weightings is an iterative process whereby each factor is rated in relation to the other in a series of pairwise comparisons (Kumaraswamy 1991). For example, the evaluation panel might begin with an estimate that minimum capital cost is worth 50 per cent on an arbitrary importance rating scale. If this scenario involved alternative design elements or was a design and construct bid then the tenderer may be required to provide estimates of cost in use or life-cycle cost (possibly assessed and verified by a construction cost consultant or facilities management consultant). Operating cost may be argued as equally important and so it could also be given a score of 50. After further discussion the construction time may be allocated a relative score of 30 (if considered to be 60 per cent as important as either operating or capital costs). In this way, an agreed total score can be developed. Simply dividing each criterion score by the summation of the scores for all criteria provides an importance index for each criterion. The total of all such indices would then be 1.00.

Table 10.3 illustrates the chosen criteria in this simple illustration, the importance indices and the hypothetical results for two of the five tenderers, A and B. The other three tenderers would of course also be rated similarly and the scores recorded as evidence of a structured and transparent evaluation process.

The results in Table 10.3 indicate that Tenderer B is more attractive than Tenderer A. Tenderer A had the lowest capital construction cost with Tenderer B being 5% more expensive. So B's rating is 1/1.05 = 0.95. Tenderer B had the lowest annual operating cost, while A was 5% higher in this respect. Both A and B did not offer the shortest construction period among those submitted. Negotiations could take place (if permissible under the specific Condiditons of Tender) to verify the durations assumed, the logic and sequencing of activities and the project method statement that affected the assumptions underpinning the quoted time. Ratings assigned for other measures reflect the evaluation panel assessments of the potential performance of the tendered submissions/tenderers.

Possible modifications and refinements

Interviews of practitioners have revealed a general reluctance to apply relatively lower importance indices to the capital cost criterion (as were applied

Table 10.3 Example of a basic comparison of two tenderers, A and B, with alternative designs

Criterion	Importance index	Rating Score			
		A	B	A	B
Capital cost	0.23	1.00	0.95	0.23	0.22
Annual operational cost	0.23	0.95	1.00	0.22	0.23
Time	0.12	0.90	0.98	0.11	0.12
Quality	0.08	0.80	1.00	0.06	0.08
Safety, health and environment	0.04	0.80	0.80	0.03	0.03
Expertise	0.08	0.60	0.80	0.05	0.06
Experience	0.04	0.80	1.00	0.03	0.04
Relationships	0.12	0.60	0.80	0.07	0.10
Community support and ethics	0.04	0.20	0.40	0.01	0.02
Claims and dispute avoidance potential	0.02	0.60	0.80	0.01	0.02
Totals	1.00			0.82	0.92

in the above example). This is particularly so in relation to construct-only contracts where cost has traditionally been the sole criterion and where rejection of the lowest tenderer usually requires elaborate justifications – expected or demanded from the evaluation panel – to avoid allegations of favouritism or corruption, for example.

A danger does in fact exist of selecting a tenderer with a significantly higher tender price if the evaluation panel were merely to follow the methodology in the above example in case the tenderer is remarkably superior to the field in most other aspects.

One approach that may be used to overcome such apprehensions and dangers in construct-only tender evaluations is to apply the proposed methodology in a second-stage exercise, that is, only after a first stage which narrows the field to those whose tender prices are in an acceptable range. This could be done by filtering out tenders that are more than 10 per cent above or 15 per cent below either the client's (or engineer's or quantity surveyor's) estimate or the average of the tenders received (if there was no detailed client estimate), for example, or by choosing an even narrower band, depending on how many are within this range and also on the client's desire for economy.

This approach is similar (but proceeds in the opposite direction) to the two-envelope system that has been used for selecting consultants, where the applicants submit their technical proposal (for executing a project and pro-viding their services, with details of staff deployed, etc.) in one envelope and their fee proposal in another sealed envelope. The envelopes containing the fee proposals are opened only in the cases of those who are shortlisted on the basis of acceptable or good technical proposals. This is useful in avoiding 'lay' client temptations to grasp at applicants who offer a low fee level and who may then provide correspondingly lower service levels or who may lead the client up a path of hidden costs.

The proposed two-stage approach in evaluating construct-only, construct with design alternative or even design and construct tenders could thus be a useful refinement that would facilitate client acceptability of the multiple criteria tenderer evaluation methodology proposed herein. This refinement may also be needed gently to shift the mindset of certain clients who have grown accustomed to the traditional and superficially economical one-dimensional cost-only criterion.

Research findings on the evaluation of some specific criteria

The needs for appropriate performance indicators for evaluating chosen criteria

The illustration of the basic model through the worked example in Table 10.3 next raises questions of how the client and its advisors should evaluate the chosen criteria and how tenderers can best present their submissions to demonstrate their strengths against these criteria. It was even found that 'bidding skills have outweighed technical expertise in winning work' based on a survey of 293 companies in the UK (*Construction Manager* 1997), with 'superbidders' four times more likely than the losers to understand the business climate of their clients, and six times more likely to understand the cost of ownership issues considered by these clients.

While comparisons against basic cost and time criteria should be relatively easy to evaluate, certain aspects such as a particular propensity for claims or time overruns will be discussed in the following subsections. It is also necessary to develop a framework of realistic indicators (measures) for evaluating past and potential performance. Such a framework of criteria and indicators for evaluating past performance has been proposed (Kumaraswamy 1991) while Tam and Harris (1996), Hatush and Skitmore (1997a, 1997b) and Chan (1996) have proposed measures for predicting future performance, despite the many variables and unknowns that would be encountered in such exercises. For example, a thread that ran through most of the research findings in this area was the underlying importance of potential teamworking, organisational arragements, communications and other human factors. For instance, Chan (1996) confirmed the existence of direct relationships between the effectiveness of the construction team leader and the speed of construction, the client's satisfaction on time, the client's satisfaction on cost, the client's satisfaction on quality, the functionality of the project, the client's overall satisfaction and the contractor's overall satisfaction. Clearly, it is important for the tenderer to demonstrate the competence of the construction team leader and the proposed construction team in general.

Structured approaches to evaluating even such volatile qualitative variables (to whatever extent possible) are thus better than none. The knowledge base that may be assembled from the many studies and experiences should enable the selection of suitable criteria and indicators for a particular scenario based on standardised frameworks and guidelines, upon which project-specific requirements, conditions and priorities may be superimposed. Benchmarking can then be attempted to facilitate realistic comparisons of past and potential performance by using such selected criteria and indicators.

Construction cost overruns, claims and disputes

Initial capital cost parameters and profiles of life-cycle costs or costs in use would, of course, be developed between designers and clients, for which a clear brief must effectively be communicated to a 'competent' design team. The competencies of the design team must thus be evaluated or predicted in the first instance. Second, an effective and economical design must be conveyed clearly, completely and accurately to prospective tenderers to minimise cost overruns caused by variations and/or other claims from contractors.

Many studies have identified 'variations' as one of the major sources of claims (for example, Yogeswaran and Kumaraswamy 1997; Chan *et al.* 1995); other findings from many countries are summarised elsewhere (Kumaraswamy 1997b). The causes of variations have in turn been explored in a few of these studies. Some variations are less avoidable or are unavoidable, for instance because of totally unexpected site conditions or project changes that could not have been reasonably foreseen. However, it has been found that many variations are from incomplete, inaccurate or ambiguous contract documents, be they bills of quantities, drawings, specifications or contract conditions. An analysis of such common causes leads to strategies to deal with them in future projects (Chan *et al.* 1995). Most strategies involve improvements at the design documentation and contract management phase and could thus also help to develop criteria and indicators for assessing design and build contractors or those who submit design alternatives. Additionally, selecting less 'claims-conscious' contractors – based on past track records – has been seen to have merit in minimising the exploitation of loopholes by some contractors, whether for unreasonable gains or to compensate for unrealistically low tender pricing.

Further analysis of significant sources of claims led to the formulation of a 'claims focus indicator' that focuses client or managerial attention in advance on potential sources of claims by contractors for additional costs and time (Yogeswaran and Kumaraswamy 1997). A total of 14 common sources of claims (as previously identified in this Hong-Kong-based study) were assessed on the basis of frequency, magnitude and avoidability of claims arising therein. The relative significance of these sources was evaluated using the claims focus indicator, thus providing a basis to prefocus client's procurement management attention on common claims sources that are potentially large, frequent and avoidable. Analysis of the root causes contributing to these sources in turn highlighted a range of issues, including teamworking, design-related and contractor-related factors. The contractor-related factors would be examined further when developing indicators for assessing potential contractors against relevant criteria (such as items 1, 2 and 10 in Table 10.2).

A 'dispute potential index' that was developed in the USA (USCII 1994) is another useful indicator that prefocuses client's procurement management attention on potential areas of common disputes. Furthermore, while disputes were seen to arise from people, the project and the process, the people variable was found to hold the key to avoiding disputes (Diekmann and Girard 1995). This finding again suggests the potential and usefulness for developing indicators for the more appropriate evaluation of contractors (and other team participants) against the relevant criteria (such as items 1, 2 and 10 in Table 10.2), also given that most disputes are in any case related to claims for extra time and/or money.

Planning, controlling and evaluating construction time performance

Well-substantiated construction time predictions and planning will not only help clients and tenderers to develop realistic time frames, time plans and programmes, but also demonstrate management competence.

Time plans may follow best practice of indicating work methods, for example through method statements and critical path plans, possibly coupled to resource scheduling and even indicating some element of project visualisation through three-dimensional computer-aided design or virtual reality demonstration of planned sequence (Retik and Hay 1994; Faraj 1996; Fisher *et al.* 1997). Further examples of demonstration models may be found at the web site address http://www.strath.ac.uk/Departments/Civeng/conman/vrtprojects.html at the University of Strathclyde in Scotland.

The importance of construction management team skills in effective planning, control and administration has been widely underscored by many investigators (for example, Ireland 1983; NEDO 1988). It has been demonstrated that the elements of 'good' project planning include familiarity not only with the use of critical path scheduling but also with the planning approach itself (Walker 1994). This includes thinking through construction methods to achieve good flow, accurate prediction of construction duration, regular monitoring of plans at appropriate intervals and a high degree of flexibility to modify plans when circumstances so require.

The results of a study of 45 Australian construction projects indicate that a well-organised construction management team can contribute significantly to the achievement of 'good' construction time performance (Walker and Sidwell 1996). A construction team's flexibility in responding to problems and opportunities is based upon sophisticated planning for control, where plans are seen as a focus for action rather than as static or rigid commands or procedures. The effective use of planning tools for simulation analysis should be encouraged so as to predict a range of possible 'futures' (future scenarios) from which a 'preferred future' can be selected. This needs to be integrated into the monitoring cycle, checking reality against the 'preferred future'.

In discussing the issue of construction time performance (CTP), an important question arises as to how CTP can be measured. A methodology for measuring CTP has been developed and tested in three recent research studies (Walker 1994; Walker and Sidwell 1996; Walker and Vines 1997). A detailed explanation of the methodology may be found elsewhere (Walker 1997b), but an introductory summary is provided in this chapter.

The concept of performance measurement envisions the comparison of an actual outcome with its expected outcome. In the case of CTP it has been found possible to derive formulae from representative samples of projects to predict expected project duration based on sets of variables that have first been found to be significant. Regression analysis has been used to derive such formulae in the above studies as well as in many others, including Bromilow *et al.* (1980) and Ireland (1983). Of course, the parameters could vary across the types of projects being considered. For example, prediction models have been developed by means of such regression analysis and significant variables have been identified for a specific group of public housing building-construction projects in Hong Kong (Chan and Kumaraswamy, in press). Artificial neural networks have also been used, either in place of regression analysis for similar predicitons (Li, 1995a, 1995b) or to supplement such analysis in predicting cost and time performance levels (Dissanayaka and Kumaraswamy 1999).

The next step is to compare the predicted construction time, as derived from the formula, with the actual construction time. A CTP index is derived from this comparison. All approved extensions of time were included in the actual construction time for projects, as these were granted to include for scope changes (Walker 1994). Using this methodology (Walker 1994), a project with an expected construction time of 110 days and an actual construction time of 100 days would have a CTP index of 110/100 (1.10), that is, it would be 10 per cent better than the 'norm'. The CTP index of a group of projects can be used as the basis of statistical analysis such as for *t* tests, analysis of variance (ANOVA) or other statistical tests that are appropriate for the sample size.

These tests can also be used to compare the difference between group CTP indices for various groups of data. An example of this is to gather data on projects in the group on the level of analysing construction methods as part of the construction planning method. The data can use a seven-point scale, for example, where 1 = very low, 2 = slightly low, 4 = average, 5 = slightly high, 6 = high and 7 = very high. Factors (as in Table 10.4) were also measured using an impact scale, where 1 = strongly diminished, 2 = diminished, 3 = slightly diminished, 4 = neither diminished nor enhanced, 5 = slightly enhanced, 6 = enhanced and 7 = strongly enhanced.

The results can be used to investigate if the mean CTP of clusters of projects in a group of data values (say, less than a value of 5 out of 7) is significantly different from values of 6 or 7 or 'greater than or equal to 5'. If there is no significant difference in the means of the clusters, then the performance characteristic is not deemed a significant factor. If there is a significant difference in the means, then the factor concerned is deemed to be a factor that is significantly associated with improved CTP. Furthermore, if the means of cluster groups are known then the extent of potential impact and a value which forms a 'trigger point' can be determined. This methodology provides a useful way in which CTP can be measured and how the results can be interpreted. Table 10.4 presents recently reported results (Walker and Sidwell 1996).

In Table 10.4 column 1 provides a description of factors shown to affect significantly (at the 95 per cent confidence level) construction time per-

Table 10.4 Construction management team factors affecting construction time performance (CTP): (a) planning, control and administration skills; (b) organisational structure and characteristics

Factors affecting CTP	CTP	Increase in mean CTP (%)
(a)		
Level of general management performance of the construction management team	⩾ High	20
Impact of construction management team's:		
management systems	Strongly enhanced	23
forecasting planning data	Strongly enhanced	31
analysing construction methods	Strongly enhanced	29
analysing resource movement	Strongly enhanced	33
monitoring and updating plans	Strongly enhanced	26
response to problems or opportunities	⩾ Enhanced	33
coordination of resources	Strongly enhanced	26
(b)		
Impact of construction management team's:		
developing an effective site team	⩾ Enhanced	23
team to manage risk	⩾ Enhanced	34
Key sub-contractor's level of:		
mechanistic style of organisational behaviour	High, very high	43
flexible style of organisational behaviour	⩾ Average	36
Construction manager's level of task-orientated management style	⩾ High	21
Site supervisor's level of:		
task-orientated management style	Very high	32
people-support-orientated management style	Slightly high, high	32
use of direct power	⩾ Average	29

formance. Column 2 provides the trigger point between two or more clusters of data tested using the analysis of variance (ANOVA) statistical test. Column 3 provides the percentage difference in CTP between the cluster of projects at or above the trigger point and the cluster below that trigger point. For example, the factor 'impact of construction management team's analysing construction methods' in Table 10.4(a) shows a 29 per cent improved CTP for projects where this impact was reported as being 'strongly enhanced'. That is, for construction management teams rated as 'very high' in this performance measure, their group of projects were on average 29 per cent better in CTP than those teams with 'high', 'average', down to 'very low' performance in this indicator.

The results in Table 10.4(a) interestingly indicate that construction management teams need to exhibit very strong skills in the relevant factors to make a significant difference to CTP. However, the rewards also appear high, being associated with increases in CTP of 20–33 per cent.

These results provide useful benchmark indicators for success. Clearly, very high levels of planning skills are associated with excellent CTP and it follows that to achieve good CTP the construction management team needs to think beyond planning as a technique and treat it as a philosophy. Indicators of 'high' and 'very high' planning skills can be measured in terms of the following two statements: 'Everybody understands their role in achieving the enterprise and project plans and objectives', and 'Continuous improvement is accepted as a strategy for survival and competitive advance' (CIDA 1993b: 134).

The particular planning skills identified in Table 10.4(a) were strongly correlated with each other (Walker 1994). They were also generally strongly correlated with communication skills and good interteam relationships. They were found to be moderately correlated with a flexible management task-orentated style. The effectiveness of the construction management team planning skills is also moderately associated with the client representative's ability to make authoritative decisions and to communicate these clearly. Such correlations between the significant factors may explain the apparently high CTP gains that were found to be associated with certain factors, as seen above.

It is interesting to note that there needs to be a good relationship between the construction management team and the client representative, or at least one in which decision-making and communication is undertaken in a manner which recognises these two teams as professionals. This underscores the importance of teamwork in achieving good CTP.

The criticality of an effective management structure in undertaking construction work has been confirmed (Morris 1994; Walker 1993). The importance of construction planning, control and administration skills have been highlighted earlier in this sub-section, as has the need for effective communication. These cannot be effectively accomplished without mobilising (selecting) 'quality' teams. The Australian Construction Industry Institute (ACII 1995a, 1995b) also stressed the importance of this 'people' factor in project management.

The correct balance of skills also needs to be available within the teams to get the work done efficiently within effective organisations. Moreover, the way in which people interact also affects the effectiveness of teams. Table 10.4(b) indicates some interesting benchmark measures that have been derived (Walker 1997a) for establishing such effective teams. These could be used as indicators when evaluating potential contractors.

An effective or highly effective construction management team organisation is evidently needed to achieve high CTP levels. For example, construction team leaders with a 'high' or 'very high' level of task orientation are associated with 21 per cent improved CTP. The important roles of subcontractors and site supervisors are also highlighted in Table 10.4(b).

Table 10.4 clearly also indicates the scope and value both for contractors and for clients to commission post-project evaluations based on such critical measures of management structure and style. These performance measures would be useful both in demonstrating and in evaluating management expertise.

Conclusions

Appropriate frameworks of multiple criteria and identified indicators for evaluating construction tenderers are evidently critical in the context of the increasingly complex requirements of multiple stakeholders. For that matter, even a one-person client commissioning a simple project would need to identify and evaluate tenderers against more than merely the apparent initial cost that may be superficially indicated by the lowest tender price. Hidden costs could arise, for example, from potential cost and time overruns, rework, quality and performance shortfalls, accidents, environmental damage and resources deployed in settling claims and disputes.

The worked example on a basic multicriteria tender evaluation demonstrates its advantages, as well as the flexibility to accommodate a range of stakeholder concerns. 'People' or 'human' factors and managerial factors such as teamworking skills and organisational structures and styles are found to be crucial. Difficulties inherent in evaluating such qualitative variables are reduced by using indicators such as those developed in relation to their impact on CTP. More development is needed following further data collection in other areas so as to improve the reliability and sensitivity of the proposed system.

The criteria or indicator framework developed for such assessments of potential contractors could also be profitably applied to evaluating the management of past projects. The considerable growth in the body of recent research findings relating to critical success factors and significant criteria suggests the potential for developing knowledge bases and benchmarking. Initiatives to develop and standardise some hybrids of the proposed frameworks could lead to a set of basic templates (with differences to cater for different project types) that may be applied to new evaluation scenarios, along with guidelines to incorporate project-specific criteria and indicators. The latter would, of course, be critical in incorporating particular project conditions, along with client requirements and priorities. Decision aids ranging from checklists to expert system front-ends based on the foregoing knowledge bases may be incorporated to assist clients and their advisors further in making the best choice of contractor.

References

Abidali, A.F. and Harris, F. (1995) 'A methodology for predicting company failure in the construction industry', *Construction Management and Economics*, 13: 189–96.

ACII (1995a) 'Benchmarking engineering and construction – review of performance and case studies', Australian Construction Industry Institute and University of South Australia, Adelaide.

ACII (1995b), 'Benchmarking engineering and construction – winning teams', Australian Construction Industry and University of South Australia, Adelaide.

Assaf, S.A., Al-Hammad, A. and Ubaid, A. (1996) 'Factors effecting construction contractors' performance', *Building Research and Information* 24(3): 159–63.

Bromilow, F.J., Hinds, M.F. and Moody, N.F. (1980) 'AIQS survey of building contract time performance', Australian Institute of Quantity Surveyors, *The Building Economist* 19(2): 79–82.

Bubshait, A.A. and Al-Gobali, K.H. (1996) 'Contractor prequalification in Saudi Arabia', *Journal of Management Engineering* (March/April): 50–4.

Chan, A.P.C. (1996) 'Determinants of project success in the construction industry of Hong Kong', The International Centre for Management and Organisational Effectiveness, Faculty of Business and Management, University of South Australia, Adelaide.

Chan, A.P.C., Yeung, C.M. and Tam, C.M. (1995) 'Effective measures for reducing variations – an Australian perspective', *Australian Institute of Building Papers* 6: 113–20.

Chan, D.W.M. and Kumaraswamy, M.M. (1995) 'A study of the factors affecting construction durations in Hong Kong', *Construction Management and Economics*, 13(4): 319–33.

Chan, D.W.M. and Kumaraswamy, M.M. (in press) 'Modelling and predicting construction durations in Hong Kong public housing', *Journal of Construction Management and Economics*.

CIC (1992) *Guidelines for the application of competitive tendering*, Construction Industry Council, London: Thomas Telford.

CIDA (1993a) 'The Australian construction industry – pre-qualification criteria for contractors and sub-contractors', Construction Industry Development Agency, Sydney.

CIDA, (1993b) 'Project performance update – a report on the time and cost performance of Australian building projects completed 1988–1993', Construction Industry Development Authority, Sydney.

CIDB (1995) 'Higher premium for certified contractors', Construction Industry Development Board, *Construction Focus* 7(3): 2.

Civil Engineers Australia (1997) 'Reduced design quality investigated', *News: Institution of Engineers Australia, Civil Edition* 69(12): 14.

Construction Manager (1997) 'Research highlights the success of superbidders', news item, Chartered Institute of Building (April): 6.

Diekmann, J.E. and Girard, J. (1995) 'Are contract disputes predictable?', *Journal of Construction Engineering and Management* 121(4): 355–63.

Dissanayaka, S.M. and Kumaraswamy, M.M. (1999) 'Comparing contributors to time and cost performance in building projects', *Building and Environment Journal*, 34(1): 31–42.

Edum-Fotwe, F.F., Price A.D.F. and Thorpe, A. (1995) 'Transformed financial ratio models for improved contractor evaluation', *1st International Conference on Construction Project Management* Nanyang Technological University, Singapore, 559–67.

European Construction Institute (1992) *Total project management of construction safety, health and environment*, London: Thomas Telford.

Faraj, I. (1996) 'The use of VR in support of construction tasks', *CIB W-65 Symposium 1996, Organisation and Management of Construction*, Glasgow: London: E & FN Spon, 142–9.

Fisher, N., Barlow, R., Garnett, N. and Newcombe, R. (1997) *Project modelling in constructiion . . . seeing is believing*, London: Thomas Telford.

Graham, P. and Smithers, G. (1996), 'Construction waste minimisation for Australian residential development', *Asia Pacific Journal of Building and Construction Management* 2(1): 14–19.

Griffiths, A. and J.D. Headley (1997) 'Using a weighted score model as an aid to selecting procurement methods for small building works', *Construction Management and Economics* 15(4): 341–8.

Hatush, Z. and Skitmore, M. (1997a) 'Criteria for contractor selection', *Construction Management and Economics* 15: 19–38.

Hatush, Z. and Skitmore, M. (1997b) 'Evaluating contractor prequalification data: selection criteria and project success factors', *Construction Management and Economics* 15: 129–47.

Herbsman, Z.J. (1995) 'A + B bidding method – hidden success story for highway construction', *Journal of Construction Engineering and Management* 121(4): 430–7.

Holt, G.D., Olomololaiye, P.O. and Harris, F.C. (1993) 'A conceptual alternative to current tendering practice', *Building Research and Information* 21(3): 167–72.

Holt, G.D., Olomololaiye, P.O. and Harris, F.C. (1994) 'Evaluating performance potential in the selection of construction contractors', *Engineering, Construction and Architectural Management* 1(1): 29–50.

Holt, G.D., Olomololaiye, P.O. and Harris, F.C. (1996) 'Tendering procedures, contractual arrangements and Latham: the contractors' view', *Engineering, Construction and Architectural Management* 3(1/2): 97–115.

Ioannou, P.G. (1993) 'Average-bid method – competitive bidding strategy', *Journal of Construction Engineering and Management* 119(1): 131–47.

Ireland, V. (1983) 'The role of managerial actions in the cost, time and quality performance of high rise commercial building projects', University of Sydney, Sydney.

Kaka, A. and Price, A.D.F. (1991) 'Relationship between value and duration of construction projects', *Construction Management and Economics* 9(4): 383–400.

Kumaraswamy, M.M. (1991) *Evaluating the management of construction projects*, PhD thesis, Loughborough University of Technology, Loughborough.

Kumaraswamy, M.M. (1996) 'Contractor evaluation and selection: a Hong Kong perspective', *Building and Environment Journal* 31(3): 273–82.

Kumaraswamy, M.M. (1997a) 'Appropriate appraisal and apportionment of megaproject risks', *Journal of Professional Issues in Engineering* 123(2): 51–6.

Kumaraswamy, M.M. (1997b) 'Conflicts, claims and disputes in construction', *Engineering, Construction and Architectural Management* 4(2): 95–112.

Kumaraswamy, M.M. (1997c) 'Repackaging construction megaprojects and redefining technolgy transfer', *Proceedings of CIB W78 Conference on Information Technology Support for Construction Process Reengineering*, Cairns, Australia, July, 215–24.

Kumaraswamy, M.M. and Dissanayaka, S.M. (1996) 'Procurement by objectives', *Journal of Construction Procurement* 2(2): 38–51.

Kumaraswamy, M.M. and Thorpe, A. (1996) 'Systematizing construction project evaluations', *Journal of Management Engineering* 12(1): 34–9.

Latham, M. (1994) *Constructing the team. Joint review of procurement and contractual agreements in the United Kingdom construction industry: final report*, London: The Stationery Office.

Li, H. (1995a) 'A neutral network-based cost estimating model – system design', *Australian Institute of Building Papers* 6: 45–51.

Li, H. (1995b) 'A neutral network-based cost estimating model – implementation and results', *Australian Institute of Building Papers* 6: 53–68.

Morris, P.W.G. (1994) *The management of projects – a new model*, London: Thomas Telford.

Nahapiet, J. and Nahapiet, H. (1985) 'The management of construction projects – case studies from the USA and UK', Chartered Institute of Building, Ascot.

NEDO (1988) 'Faster building for commerce', National Economic Development Office, Commercial Building Steering Group, London.

Ng, S.T., Smith, N.J. and Skitmore, R.M. (1995) *Case-based reasoning for contractor prequalification – a feasibility study. Developments in artifical intelligence for civil and structural engineering*, Edinburgh: Civil-Comp Press.

Osborne, R. (1996) 'Learning to say NO', *Consulting Engineer International* (4): 14.

Potter, K.J. and Sanvido, V. (1995) 'Implementing a design/build prequalification system', *Journal of Management Engineering* 13(3): 30–4.

Retik, A. and Hay, R. (1994) 'Visual simulation using VR', *10th ARCOM Conference*, Vol. 2, Loughborough, Association of Researchers in Construction, ARCOM, The University of Salford, Salford, 537–96.

Russel, J.S. and Skibniewski, M.J. (1990) 'Qualifier-2: knowledge based system for contractor prequalification'. *Journal of Construction Engineering and Management* 116(1): 157–71.

Russel, J.S., Hancher, D.E. and Skibniewski, M.J. (1992) 'Contractor prequalification data for construction owners', *Construction Management and Economics* 10(2): 117–35.

Songer, A.D., Molenaar, K.R. and Robinson, G.D. (1996) 'Selection factors and success criteria for design–build in the USA and UK', *Journal of Construction Procurement* 2(2): 69–82.

Stout, L.B., (1995) 'Is competitive price bidding for professional services ethical? Another view', *Journal of Professional Issues in Engineering Education and Practice* 121(4): 256–8.

Tam, C.M. and Harris, F. (1996) 'Model for assessing building contractors' project performance', *Engineering, Construction and Architectural Management* 3(3): 163–86.

Tarricone, P. (1993), 'Deliverance', *Civil Engineering* (February): 36–9.

USCII (1994) 'The continental divide of dispute resolution', US Construction Industry Institute, *CII News* VII(5): 1, 3.

Walker, A. (1993) *Project management in construction*, Oxford: Basil Blackwell.

Walker, D.H.T. (1994) 'An investigation into factors that determine building construction time performance', Department of Building and Construction Economics, Royal Melbourne Institute of Technology, Melbourne.

Walker, D.H.T. (1997a) 'Construction time performance and traditional versus non-traditional procurement systems', *Journal of Construction Procurement* 3(1): 42–55.

Walker, D.H.T. (1997b) 'Choosing an appropriate research methodology', *Construction Management and Economics* 15(2): 149–59.

Walker, D.H.T. and Sidwell, A.C. (1996) 'Benchmarking engineering and construction – a manual for benchmarking construction time performance', Australian Construction Industry Institute and University of South Australia, Adelaide.

Walker, D.H.T. and Vines, M.W. (1997) 'Construction time performance in multi-unit residential construction – insights into the role of procurement methods', *13th Annual ARCOM Conference*, vol. 1, Kings College, Cambridge, Association of Researchers in Construction Management, 93–101.

Works Branch (1997) 'Rules for the administration of the list of approved contractors for public works', works branch technical circular 9/97, Works Bureau, Hong Kong Special Administrative Region.

Yng, L.Y., Ofori, G. and Pheng, L.S. (1997) 'Developing a model for selection of consultants by design and building contractors: a pilot study', *First International Conference on 'Construction Development'*, vol. 1, Singapore, December, 374–81.

Yogeswaran, K. and Kumaraswamy, M.M. (1997) 'Perceived sources and causes of construction claims', *Journal of Construction Procurement* 3(3): 3–26.

11 Applying partnering in the supply chain

Jason Matthews

Introduction

The concept of partnering as currently practiced is not a recent pheno-menon. It has existed in various guises for many years, although not referred to by that specific name (Loraine 1994). Since the beginning of the 1990s there has been a proliferation of books, papers and manuals pertaining to partnering, all advocating their own philosophies, methodologies, theories and best practices. Partnering has now become an established concept for procuring buildings with wide acceptance in the USA, Australia and the UK.

Loraine (1994) believed that the modern thrust of partnering relation-ships originated in the associations forged between manufacturers and sup-pliers in the Japanese car industry of the 1960s and 1970s. Contrary to Loraine, the National Economic Development Office (NEDO 1991) sug-gested that partnering appeared to have evolved rather than having begun life as the realisation of a specific idea. NEDO believed that its precursors were probably much looser arrangements, such as strategic alliances or preferred-supplier associations between client and contractor where trust and confidence in each other was built up over the years. In the UK the first relationship that may be classified as a 'partnering relationship' was that of Marks & Spencer and Bovis, which commenced in the 1920s (NEDO 1991).

A recent study by Walter (1998) revealed how important partnering has become to UK construction companies. Walter reported that many firms expected to see partnering account for over two thirds of their workloads in the coming financial year (1998/9). Companies such as AMEC, Amey Construction, Bovis, HBG Construction and Alfred McAlpine Construction all stated that the majority of their current contracts were partnered. Moreover, the situation in the USA seems to be similar. Maloney (1997) reported that the use of partnering in the USA was spreading and would continue to spread. Maloney concluded that although no up-to-date data were available on the use of partnering, anecdotal evidence from industry indicated the increased use of partnering.

However, many practitioners are still not convinced that partnering can offer benefits to them. It is not uncommon for practitioners to ask, 'What benefit will it give me?' or 'What are partnering's potential risks and problems?'. The overall aim of this chapter is to seek answers to these questions and also to outline how partnering might be implemented between main contractors and subcontractors as it is this link in the supply chain which has regularly been overlooked. In order to achieve this the published literature relating to partnering has been synthesised, producing a valuable reference point for practitioners and academics.

The chapter begins with a brief overview on the development of partnering. This gives the chapter a basis on which to carry out further discussions of the definitions given to partnering and how it is categorised. This is then followed by a discussion of those benefits that can be achieved by: project owners (clients); main contractors; the architect, engineer and consultant; and subcontractors and suppliers. An analysis then takes place of the potential risks and problems of partnering. Finally, the chapter describes how partnering might be implemented, with reference to an approach developed in the UK between main contractors and subcontractors.

Defining partnering

Infante (1995) stated that defining partnering assists in understanding the concept. Crowley and Karim (1995) and I (Matthews 1996) reported that partnering is typically defined in one of two ways: first, by its 'attributes', such as trust, shared vision and long-term commitments; and, second, by the 'process', where partnering is seen as a verb and as such includes developing a mission statement, agreeing goals and organising and conducting partnering workshops. Defining partnering in these ways illustrates the intended results of partnering as well as the process that was employed to achieve these results. However, this definitional bias leaves the entity of partnering, and the partnering organisation, undefined (Crowley and Karim 1995).

Crowley and Karim (1995) state that partnering is a decentralised, pseudo-organisational structure, designed to allow better flexibility in meeting specific project needs. This organisation provides the scope to solve day-to-day problems, resolve conflicts, expedite decision-making and increase organisational competence in achieving project goals. They developed a model to depict their ideas (Figure 11.1).

Crowley and Karim's model highlights another way of defining partnering. Although they discuss theoretical issues and the work is largely theoretically based there is, they say, significant benefit to be gained by industry in improving its understanding of how partnering affects organisational interfaces.

The most commonly cited definitions for partnering are those proposed by the Construction Industry Institute (CII 1991) in its report 'In search of partnering excellence' and that put forward by the Associated General Contractors of America (AGC 1991). The CII report is based on 27

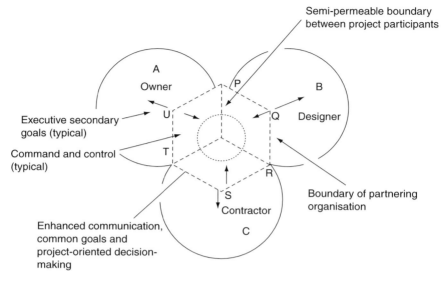

Figure 11.1 Conceptual model of partnering
Source: Crowley and Karim 1995
Reproduced by kind permission of the American Society of Civil Engineers

partnering case studies in the USA and presents a comprehensive review of the partnering subject.

The CII defines partnering as:

> A long-term commitment between two or more organisations for the purpose of achieving specific business objectives by maximising the effectiveness of each participant's resources. This requires changing traditional relationships to a shared culture without regard to organisational boundaries. The relationship is based on trust, dedication to common goals, and an understanding of each other's individual expectations and values. Expected benefits include improved efficiency and cost effectiveness, increased opportunity for innovation, and the continuous improvement of quality products and services.
>
> (CII 1991: IV)

The Associated General Contractors of America define partnering as:

> a way of achieving an optimum relationship between a customer and a supplier. It is a method of doing business in which a person's word is his or her bond and where people accept responsibility for their actions. Partnering is not a business contract but a recognition that every contract includes an implied covenant of good faith.
>
> (AGC 1991: 2)

The CII's definition of partnering covers the attribute and process definition whereas the AGC's has a strong attribute bias. An interesting feature of partnering is the difference in the CII and AGC definitions. The CII definition describes 'strategic partnering' whereas the AGC describes 'project partnering'.

Categorising partnering

Different categories of partnering relationships exist. Perhaps the most common categorisations are 'project partnering' and 'strategic partnering'. Put simply, project partnering is partnering undertaken on a single project. At the end of the project the partnering relationship is terminated and another relationship is commenced on the next project. Strategic partnering takes place where two or more firms use partnering on a long-term basis to undertake more than one construction project.

It is generally accepted that strategic partnering relationships accrue benefits on individual projects, but the scale of the benefits increase as each project undertaken profits from the lessons learnt from previous projects. Strategic partnering provides the full benefits of partnering because it allows time for continuous improvement (RCF 1995).

Benefits of applying partnering: stakeholders

Before discussing the relative benefits and potential problems of partnering it is important to identify the main stakeholders who may be involved in the partnering process. This will enable specific benefits to be attributed to each stakeholder.

Stakeholders of any partnering process are those parties, individuals or organisations who have assigned their commitment to the partnering process by signing a charter. The six primary stakeholders in a partnering project are (Matthews 1996):

- the building owner;
- the design team;
- the main contractor;
- the specialist contractors;
- the sub(sub)contractors;
- the suppliers.

The following could be added to the list depending on the type of project: relevant authorities, relevant public interest groups and the partnering facilitator. However, when identifying the benefits that can be achieved by the partnering most authors who actually discuss partnering benefits usually concentrate on some, if not all, of the following: project owner, project contractor, project architect, the engineer or consultant; project subcontractors and suppliers (Uher 1994; AGC 1991; CII 1991; Hellard 1995; RCF 1995; Stevenson 1996; Schultzel and Unruh 1996; Kubal 1994). For the purposes of this chapter those stakeholders put forward by the previously cited authors will be used.

Identification of benefits

In order to identify the benefits of partnering, under the previously stated stakeholders, an examination of the partnering literature was undertaken to identify those theorists and practitioners who had discussed the benefits in-depth. A total of 11 sources of information were found (Table 11.1). Each benefit was identified and situated under its relative stakeholder. In total, 94 benefits were identified. Very little attempt was made by the authors to attribute benefits to strategic or project partnering. Therefore, no attempt has been made here to attribute benefits under different types of partnering – benefits are attributed only under the heading of their relative stakeholders. The following describe those benefits found to be most common for each stakeholder.

The most prevalent benefits for the project owner or client are as follows:

- reduced exposure to litigation through open communication and issue-resolution strategies (eight similar citings);
- lower risk of cost overruns and delays because of better time and cost control over projects (eight similar citings);
- open communication and unaltered information allow more efficient resolution of problems (seven similar citings).

The most prevalent benefits for the main contractor are as follows:

- reduced exposure to litigation through communication and issue-resolution strategies (eight similar citings);
- better time and cost control over project (eight similar citings);
- increased opportunity for financially successful project because of non-adversarial win–win attitude (six similar citings).

The most prevalent benefits for the architect, engineer and consultants are as follows:

- reduced exposure to litigation through communication and issue-resolution strategies (seven similar citings);
- enhanced role in decision-making, as an active team member in providing interpretation of design intent and solutions to problems (six similar citings);

Table 11.1 The benefits of partnering

Benefits to project owner (Client):

1 Reduced exposure to litigation through open communication and issue-resolution strategies[a–h]

2 Lower risk of cost overruns and delays because of better time and cost control over projects[b,c,e–h,j]

3 Better quality product because energies are focused on the ultimate goal and not misdirected to adversarial concerns[b–e,j,k]

4 Potential to expedite project through efficient implementation of the contract[b,c,k]

5 Open communication and unaltered information allow more efficient resolution of problems[a–c,e,g,h,k]

6 Lower administrative costs because of elimination of defensive case building[a–c,e,k]

7 Increased opportunity for innovation through open communication and the element of trust, especially in the development of value engineering changes and constructability improvements[a–c,e,k]

8 Increased opportunity for a financially successful project because of non-adversarial win–win attitude[b,e,k]

9 Reduced risk of pressure groups disputing the project through the involvement of end users and other stakeholders[b]

10 Increased opportunity helps customers get project teams to focus their needs[h]

11 Better quality product through reduction of defects and rework[f,h,j]

12 Established team finds time to develop techniques that provide better and more relevant information[h,j]

13 Partnering helps reduce client's staff time by avoiding going through the same learning curve[d,g,h,j]

14 Partnering reduces costly and time-consuming selection processes leading to shorter design times, quicker start on-site and shorter construction times[d,h,j]

15 Partnering firms are more responsive to customer short-term emergencies and to changing project or business needs[d,h]

16 Improved understanding promotes improved safety[h]

17 Increased market share and profitability[j]

18 Improved ability to respond to changing market conditions[j]

19 Affiliation with strong and respected national firms[j]

20 Better schedule control[f]

21 Development of team for future projects[f]

22 Reduced dependence on legal counsel[f]

23 Eliminates need to allocate highly skilled resources to the bid adjudication process[g]

Table 11.1 Continued

Benefits to Contractor:

1 Reduced exposure to litigation through communication and issue-resolution strategies[a–f,h,k]

2 Increased productivity because of elimination of defensive case building[a–c,e,k]

3 Expedited decision making with issue resolution strategies[b,c,e,f,i,k]

4 Better time and cost control over project[a–c,e,f,h,i,k]

5 Lower risk of cost overruns and delays because of better time and cost control over project[a–c,e,h,k]

6 Increased opportunity for financially successful project because of non-adversarial win–win attitude[a–c,e,f,k]

7 Better quality management through a culture and organisational structure conducive to total quality management implementation[d,i,k]

8 Better integration and materials management through improved communication and team approach[g,k]

9 Quicker response time to questions and answers and processing of variations[k]

10 Opportunity to increase margin through cost sharing of post-tender value management savings[k]

11 Improved industrial safety through team approach and subcontractor involvement in planning and monitoring[d,k]

12 Opportunity to develop an ethical and mutually rewarding relationship with subcontractors[k]

13 Lower overhead costs[e]

14 Enhanced repeat business opportunity[d–h,j]

15 Increased productivity reduces direct costs[h,j]

16 Fewer person hours because of increased productivity[g,h,j]

17 Less paper[h]

18 Lower marketing costs[h]

19 Lower fees[h]

20 Allows for continuity of personnel from project to project[h]

21 Costs associated with bidding on a project basis are eliminated and in many cases replaced by a simple, essentially administrative, calculation of project sum[g,h,j]

22 Partnering reduces costly and time-consuming selection processes leading to shorter design times, quick start on-site and shorter construction times[h]

23 Contractor's increased understanding improves commissioning and handover of the building is smoother[h]

24 Understanding the customer plan allows the contractor to plan the development of its own staff more confidently in changing market conditions[h]

25 Earlier involvement of contractor, plus value engineering (conceptual phase) improves constructability[f,h]

Table 11.1 Continued

26 Improved understanding promotes improved safety[h]

27 Encourages strategic relationship-building (long-term)[i]

28 Affiliation with strong and respected national firms known in the business world[g,j]

29 Compensation based on value-added contributions[j]

30 Concentration on the continuous improvement of work systems provides greater efficiency and enhancement of technical matters[j]

31 Better allocation of resources because of ability to plan on a long-range basis[g,j]

32 Faster payments[j]

33 Formation of teams for future projects[f]

34 Improvements in operation systems which spill over into all activities[g]

35 There should be some sharing of the cost of problem solving[g]

36 Opportunity to refine and develop new skills[d]

Benefits to architect, engineer and consultants:

1 Reduced exposure to litigation through communication and issue-resolution strategies[b–f,h,k]

2 Minimised exposure to liability for document deficiencies through early identification of problems, continuous evaluation, and cooperative, prompt resolution which can minimise cost impact[b,c,e,h]

3 Enhanced role in decision-making process, as an active team member in providing interpretation of design intent and solutions to problems[b,c,e,f,h,k]

4 Reduced administration costs because of elimination of defensive case building and avoidance of claim administration and defence costs[b,c,e,k]

5 Increased opportunity for a financially successful project because of non-adversarial win–win attitude[b,c,e,f,k]

6 Less paper[h]

7 Lower marketing costs[h]

8 Lower fees[h]

9 Enhanced opportunity for repeat business[d,f,h]

10 Designers develop an understanding of each other's approach and so propose and develop designs in ways that are in tune with each other's thinking[h]

11 Partnering reduces costly and time-consuming selection processes leading to shorter design times[h]

12 Improved understanding promotes safety[h]

13 Optimum use of designers time[f]

14 Development of team for future projects[f]

15 Opportunity to develop and refine new skills[d]

Table 11.1 Continued

Benefits to Subcontractors & Suppliers:

1 Reduced exposure to litigation through communication and issue-resolution strategies[b,c,e,h,k]

2 Equity involvement in project increases opportunity for innovation and implementation of value engineering work[b,c,e,k]

3 Potential to improve cash flow owing to fewer disputes and withheld payments[b,c,e,k]

4 Improved decision-making avoids costly claims and saves time and money[b,c,e,f,h,k]

5 Enhanced role in decision-making process as an active team member[b,c,f]

6 Increased opportunity for a financially successful project because of non-adversarial win–win attitude[b,c,e,f,k]

7 Reduced delays through project team involvement in Industrial and safety issues[k]

8 Reduced exposure to bid shopping and onerous subcontract conditions[f,k]

9 Reduced overhead costs[e]

10 Better or more reliable programming[e]

11 Enhanced opportunity for repeat business[e,f,h]

12 Less paper[h]

13 Increased productivity reduces direct costs[h]

14 Fewer person hours because of increased productivity[h]

15 Lower marketing costs[h]

16 Costs associated with bidding on a project basis are eliminated and in many cases replaced by a simple, essentially administrative calculation of project sum[h]

17 Partnering reduces costly and time consuming selection processes leading to shorter design times, quick start on-site and shorter construction times[h]

18 Improved understanding promotes improved safety[h]

19 Faster payment[f]

20 Reduced dependence on legal counsel[f]

By-products of partnering:

1 Addressing human elements to build team environment, stakeholders find themselves in a new mode of thinking[b,c,e,h]

2 Working can become fun[b,c]

3 Morale is enhanced and an *esprit de corps* is developed[b,c,h,j]

4 Heightened awareness of the value of fair dealing can be used internally, externally and in all aspects of business life[b,c]

5 Demonstration of integrity and fair dealing produces respect for others. This respect produces a reputation of true value in the industry[b]

Table 11.1 Continued

6 Partnering process empowers personnel of all stakeholders with the freedom and authority to accept responsibility[e]

7 Partnering arrangements provide access to specialist bodies of knowledge owned by a partner that might otherwise not be known about or which would be difficult to access[h]

8 Partnering helps firms to become leaders in their business by innovating and exploring many attractive options[h]

9 New opportunities and career paths for employees[c,g,j]

10 Work with some sense of job security[g,j]

11 A stated and strong commitment to training and learning, both internally and externally[j]

12 Opportunity to make strong personal relationships[j]

13 An opportunity for immediate recognition and feedback, and in some cases incentive monetary rewards[j]

Sources: [a]Abudayyeh 1994; [b]AGC 1991; [c]CIDA 1993; [d]Cook and Hancher 1990; [e]Hellard 1995; [f]Kubal 1994; [g]NEDO 1991; [h]RCF 1995; [i]Sanders and Moore 1992; [j]Schultzel and Unruh 1996; [k]Uher 1994

- increased opportunity for a financially successful project because of non-adversarial win–win attitude (five similar citings).

The most prevalent benefits for subcontractors and suppliers are as follows:

- improved decision-making avoids costly claims and saves time and money (six citings);
- reduced exposure to litigation through communication and issue-resolution strategies (five citings);
- increased opportunity for a financially successful project because of non-adversarial win–win attitude (five citings).

Along with the above benefits of partnering, by-products of partnering are seen to be:

- addressing human elements to build a team environment;
- stakeholders finding themselves in a mood of thinking;
- morale being enhanced and an *esprit de corps* being developed;
- new opportunities and career paths for employees.

The main weakness in the preceding benefits is that on the whole they are intangible. However, both the Arizona Department of Transportation

(ADOT) and the United States Army Corps (USAC) have reported significant tangible benefits of implementing project partnering.

The ADOT reported that 120 partnered projects achieved the following benefits: an average time saving of 12 per cent; cost savings of 18 per cent (percentage of bid amount); value engineering savings of 3 per cent; and total cost savings of 20 per cent (percentage of bid amount). These results are supported by the benefits achieved by the USAC. USAC reported that 16 partnered projects produced a mean cost change on partnered projects of +3 per cent, 6 per cent lower than that of non-partnered projects. Also, variations on partnered projects were 4 per cent lower than those on non-partnered projects. However, more significantly, claims costs were 4 per cent lower than on non-partnered projects (Weston and Gibson 1993).

A connection between these two sets of results lies in the fact that all of the projects in question were undertaken within the public sector (in the USA). Although it can be argued that this may not have any significance on the overall benefits gained, data do exist to support the notion that private sector projects can also receive benefit from adopting a partnering approach.

The Reading Construction Forum (RCF 1995) in the UK quote that strategic partnering can achieve savings of 30 per cent over time and that project partnering can achieve savings of between 2–10 per cent. My own research (Matthews 1996) on the effect partnering has on the main contractor–subcontractor relationship identified that subcontractors quote approximately 10 per cent lower on partnered projects than they do on non-partnered projects.

Identification of potential risks and problems of partnering

Despite the potential advantages derived from partnering arrangements, partnering entails risks and problems. In order to identify these, an analysis of partnering literature was undertaken. In total 13 authors who discussed risks and problems of partnering were identified. Those statements put forward by the authors are included in Table 11.2. From Table 11.2, the following observations can be made.

Commitment by all parties can be seen not only as a very prevalent problem in partnering, but also to take various forms, including commitment in finding work, uneven levels of commitment, continuity of commitment, lack of commitment and up-front commitment.

Cultural issues of changing the way a company works can also be seen as important. Adversarial ways run deep, and are hard to change. A start may be made under partnering philosophy but this may change back to a traditional philosophy (a problem identified and supported by six studies). Moreover, this problem also had connecting implications for other risks identified. Statements 2, 7, 14, 15, 23, 25, 27 and 29 in Table 11.2 can all be seen as by-products of this change back to traditional habits.

Table 11.2 Identified potential risks or problems of partnering

1	Too many people jump on the bandwagon[b]
2	A loss of control or even dishonesty arising from relaxing the congenital contract[b,l]
3	The adoption of partnering to correct weakness in an organisation[b]
4	It takes time to achieve properly, up-front investment; commitment is needed[h]
5	Adversarial ways run deep and are hard to change. A start with partnering philosophy may revert back to traditional philosophy[a,e,f,h,l,m]
6	Uneven levels of commitment[d,f]
7	Failure to share information[f]
8	Continual commitment to partnering is needed[f,m]
9	Lack of commitment[j,m]
10	Dilution of the impact of partnering (conservatism)[j]
11	Over optimism when benefits will be achieved[l]
12	Arrangement goes soft and the benefits dissipate[j]
13	Selection of wrong partner[j]
14	The partnership is not treated as an equal partnership[j]
15	Creation of master–servant relationship[j]
16	Lack of provision of continuity of workload[j]
17	Closer relations promote unethical collusion[d,i]
18	Increasing amount of time spent is in meetings[k]
19	Unfulfilled expectations[e]
20	One size (process) fits all solutions[e]
21	Lip service is given to partnering[a]
22	Loss of opportunities for contractors[g]
23	Lower margins for contractor[g]
24	Variable workload will make commitment more onerous[g]
25	Absence of competitive bidding may reduce benefit gained by market downturns[g,l]
26	Committed to find work for core team[g]
27	Difficulty in keeping contractors on their toes[g]
28	Adverse effect on staff[g]
29	Reduction in opportunity to benefit from client mistakes[g]
30	Charter and contract not compatible[m]
31	Evaluation and assurance of value achieved[l]
32	Creation of dependencies[l]
33	Internal concerns about job security[l]
34	Increased stress resulting from higher client expectations[l]

Sources: [a]AGC 1991; [b]CIBSE 1995; [c]CII 1991; [d]Cook and Hancher 1990; [e]Harbuck *et al.*, 1994; [f]Moore *et al.* 1992; [g]NEDO 1991; [h]Pakora and Hastings 1995; [i]Plavsic 1994; [j]Porter 1996; [k]Sanders and Moore 1992; [l]Schultel and Unruh; 1996; [m]Uher 1994

Using partnering as a market or promotional ploy or not using it properly can also be identified as a problem or risk. Statements 1, 3, 10, 11, 12, 13, 19, 20 and 21 all support this.

Other notable correlations include the legal implications of comments 2 and 30, the careful and realistic evaluation of the benefits of partnering (statements 31 and 11) and the appearance of unethical practices as identified (statements 2 and 17).

Semi-project-partnering approach

Background

A major UK contractor realised in 1990 that the construction industry could no longer continue to employ the traditional adversarial approach to working with subcontractors. The main contractor believed that in order to work more productively it had to work more closely with its subcontractors and develop closer working relationships. The main contractor developed a three-stage research strategy (see Matthews *et al.* 1996) and set about defining and developing new working relationships based on partnering. The partnering approach described in this chapter has been extensively adopted throughout the UK by the main contractor who developed it. Although the approach is primarily applicable to the main contractor–subcontractor relationship the methods used in the semi-project-partnering approach can be adopted throughout the supply chain. The approach was termed 'semi-project-partnering' because it was understood by the main contractor that 'true' partnering would be based on negotiation rather than competition. The semi-project-partnering approach employs limited competition using the fundamental principles of project partnering.

The semi-project-partnering approach has three main phases:

- procurement set up: package and company identification;
- initial selection and notification: first subcontractor interview and second subcontractor interview;
- selection and appointment of subcontractor: third interview and tender clarification, subcontractor selection, project day and pricing.

Figure 11.2 shows diagrammatically the sequence of events within the implemented semi-project-partnering approach.

Project details

The first project that the semi-project-partnering approach was implemented upon was a commercial development, located in the UK Midlands, with a contract value of £14.5 million. The development had already been let before construction commenced, with the eventual tenant becoming a crucial member of the partnering team.

The project team.

A project team was established for the duration of the project. Although it was understood that the personnel on the team would change during the life of the project, members of the team during the partnering stages were: the main contractor's commercial representatives (estimator); design repre-

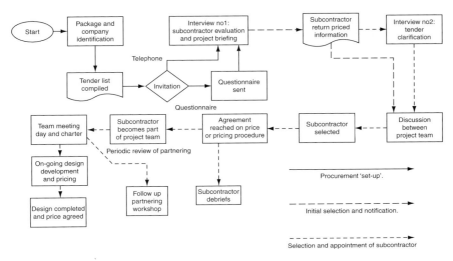

Figure 11.2 Implemented semi-project-partnering approach
Source: Matthews 1996

sentatives (architect and project design manager); engineering representatives (structural and services); and site management representatives (project manager and contracts manager). The client representation included: the client himself; the private quantity surveyor; the architect; and engineering representatives (structural and services). The project team met at least once a week in order that progress and problems could be discussed.

Stage 1: package and company identification

The objective of this stage was to identify all major packages on the project that would accrue benefit from utilising the partnering approach. Packages were examined under the following headings:

- design content;
- complexity of construction;
- high subcontract value;
- long periods of construction;
- early commencement of construction;
- high levels of aesthetics;
- long procurement times;
- trades that could 'add value' with their early input.

Having identified all trades or packages, the names of subcontractors (SCs) were put forward by all members of the project team during a weekly team meeting. Each team member, including the main contractor's site staff, had an opportunity to comment on the names put forward. The team assessed each company in order to shorten the list. SCs were assessed primarily on their past performance. Other criteria employed included:

- ability to produce the required standard of work;
- ability to undertake the work (not wanting to 'overstretch' the subcontractors);
- positive attitude (past experiences);
- firm financial background;
- good in-house design service (where applicable);
- good standards of management (site and head office);
- consideration of whether the main contractor would want to develop a long-term relationship with the SC?

On conclusion of the team meeting, between three and five SCs remained for each package. It was then agreed by all team members that these SCs could offer a comparable service.

A member of the main contractor's project team, usually the estimator, then contacted each of the SCs and invited them to an interview. The team member informed the SC of the type of project, its location and programme time, the identity of client and when work was likely to begin. A confirmation letter then followed up the invitation.

Stage 2: first subcontractor interview

The first subcontractor interview had three aims: first, to assess the SC's ability in terms of attitude, proactiveness, design capability, honesty, background and work load; second, to introduce the project and the partnering philosophy to the SCs; third, if available, to hand over pricing details and other relevant information concerning that package.

The project team was represented by: the private quantity surveyor; the project estimator; the project manager (for primarily construction trades); the appropriate engineer (services or structural trades); the project design manager and in certain instances the architect for those trades with high aesthetical values; and a planner for those trades with programming implications.

The project design manager acted as the chairman for the meeting. The SCs were told under what criteria their submissions were to be evaluated. These criteria included:

- an understanding of the partnering concept;
- response to partnering;
- alternative proposals put forward for their package, including design innovation, alternative product specification and value engineering;

- indicative price;
- technical ability;
- enthusiasm for the project;
- past experience of similar work.

During this interview all SCs were supplied with a maximum cost plan price and general arrangements, sections, specifications and approximate quantities.

At the end of the meeting each project member assessed the SCs using a standard pro forma. The project estimator kept the pro formas. The project estimator informed those SCs who did not reach the requirements of the project team that they would not be required to tender on the project.

Stage 3

TENDER CLARIFICATION

This interview was used when the first interview acted as an initial SC evaluation and the second was used to hand over pricing documentation. The aim of this interview was to give the project team the opportunity to discuss the tender and check for compliance and accuracy following the return of the tender documents. The client's project manager and quantity surveyor, project estimator, project engineer and the project design manager attended this meeting.

SUBCONTRACTOR SELECTION

The final decision on which SC was to be used was left primarily to the main contractor, although the client's quantity surveyor did have an input into the decision.

The selection was made based on those criteria communicated to the subcontractor during the first interview. The selections were made at a meeting which was attended by the project design manager, project estimator, main contractor's project manager, project buyer, client's quantity surveyor and the client's project manager.

PROJECT DAY

A one-day project workshop was initiated and paid for by the client. The day was held at a hotel in a neutral location away from the normal meeting areas. The aim of the day was to bring together all 'partnering' members of the project team, to review the progress made to date, where progress needed to be made in the future and to sign the partnering charter. All parties that had worked on the project were represented at the project day.

These included the structural and services engineers, concept architects, tenant, main contractor, project managers and all partnering SCs.

The day commenced with the client and the main contractor reviewing the progress made to date on the project. Each partnering party then presented their requirements or objectives for the project and what they could bring to the project. Once the presentations were completed the partnering charter was agreed upon and signed.

The partnering charter

A new approach in developing a partnering charter was devised (Figure 11.3). It was noted that the partnering literature (RCF 1995; Cowan and Gray 1992; Mosley *et al.* 1991) advocated that the development and the format of a partnering charter should contain the following qualities:

- the actual process should be planned out;
- the charter needs to be prepared by all, not just by a committee;
- the charter size should be limited to one page;
- the charter should, where possible, be multicoloured;
- all key logos of organisations should be incorporated where possible;
- there should be no individual titles or signature blocks;
- the charter should be signed by all that participated in its development.

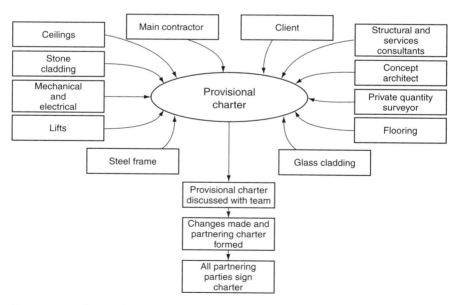

Figure 11.3 Charter formulation process
Source: Matthews 1996

Before the project day each party was asked to identify 10 objectives that they wanted to achieve during the project. It was suggested that each party establish its own individual company meeting where different members of staff could have their input. Once all objectives had been returned the most prevalent objectives were incorporated into a provisional project charter.

During the project day the provisional charter was presented to the attendees. The charter was presented via a computer link to a personal computer, which enabled changes to be made to the charter as they were agreed. All parties were asked if they approved of the content of the charter, changes were made and the final charter was printed out. Once the charter had been checked all attendees signed the charter to signify their commitment to it. Copies of the charter were distributed to all attendees on the project day.

Issue resolution policy

An issue resolution policy (IRP) was developed and implemented (Figure 11.4) in order that any grievances could be resolved at the lowest level of management. All parties on the project were aware of the IRP. The following fundamental rules and guidelines were identified in the partnering literature (RCF 1995; Cowan *et al.* 1992; AGC 1991) and were included in the IRP:

- IRP should be developed during the workshop;
- three separate levels (technical, managerial and political) of problem solving should be established;
- a problem-solving team should be established at each level with the aim being to bring partnering parties together;
- each level should have a time limit;
- any partnering party may raise an issue;
- all problems should be dealt with first at the lowest level; if the matter is not agreed in a specified time period then it should be moved to the next level;
- no jumping of levels should be allowed;
- ignoring the problem or 'no decision' is not acceptable;
- if the dispute is not settled in three stages then mediation may be employed.

An IRP committee was formed that acted as the problem-solving team. The committee members were selected and included: four members from the client's team; the main contractor's quantity surveyor, estimator, project manager, contracts manager and planner; two facilitators and one representative from each of the partnering subcontractors.

The IRP had the following six primary rules:

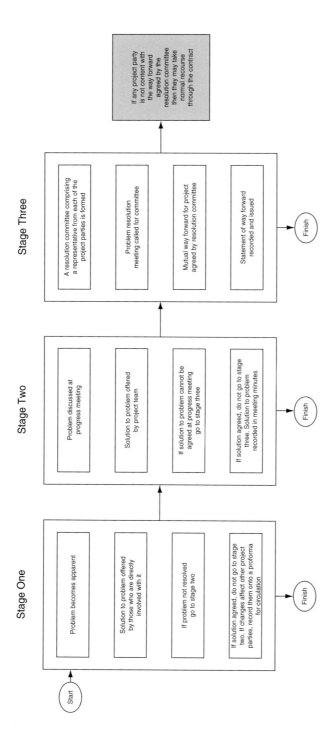

Figure 11.4 Issue-resolution policy
Source: Matthews 1996

- each level of resolution had a 24-hour time limit until the issue is escalated to the next level;
- all issues were to commence at level 1 and every effort was to be made to resolve problems at the lowest level;
- any partnering party could raise an issue;
- no jumping of levels was permitted;
- failure to reach a decision was unacceptable;
- the issue-solving team had 24 hours notice to form.

Partnering evaluation process

A partnering evaluation process was developed which evaluated the impact of the partnering during the course of the project.

Questionnaires were completed monthly and forums were carried out after key stages of the project had been completed. These forums gave the project members a 'platform' to voice their opinions on the strengths and weaknesses of the project. The partnering evaluation form contained seven 'tick-the-box' questions. Each question dealt with an objective from the partnering charter. Each question had to be ranked from 1 (very poor performance) to 4 (excellent). Any respondent answering below 2 was requested to give reasons within the comments section. The evaluation form also contained two open-ended questions. These questions dealt with how partnering expectations were being met and requested any suggestions that may improve the partnering on the project.

Debrief unsuccessful subcontractors

Each unsuccessful subcontractor was asked if they would like to attend a debrief and identify why they were not awarded the work package – all subcontractors accepted the offer. The debrief was chaired by the project design manager and was attended by the project buyer and either an architect or engineer depending on what trade was being debriefed.

Follow-up workshops

The aim of the follow-up workshop was to reinforce the team-building developed at the initial workshop and to assess the progress of the project. In doing this the partnering workshop enabled:

- renewal of partnering commitment;
- building of teams;
- review of project performance;
- further development of action plans to improve the performance of all work processes;
- introducing new project staff and SCs to the partnering process.

The workshops:

- were held at three-month intervals;
- were scheduled for one day;
- employed an external facilitator;
- involved team-building exercises to promote synergy.

The purposes of the workshops were threefold: first, to re-emphasise the partnering concept, including the charter, to the original participants who had attended the initial project day: second, to inform new subcontractors joining the project about partnering and to demonstrate team working on the project; finally, in a fun way, to reinforce the sense of team working and make explicit the feeling towards the partnering concept.

The objectives were realised in a number of group sessions based on the following themes:

- depicting diagrammatically typical feelings and relationships on a traditional project and also on a partnering project in order to identify the differences;
- using the characteristics of the 'utopian' partnering project as a basis for evaluating performance of the project;
- generating ideas for improvement for areas given low grades during the evaluation.

After the completion of each workshop the facilitators produced a synopsis of the day. The synopsis acted as a reminder of the issues that the project team believed were important. The following are three points that were made during one of the workshops:

- the level of commitment and involvement to the concept of partnering had still to reach some of the operatives on the project;
- some parties had not been involved at an early enough stage in the design process, or their involvement had not been continuous;
- the main contractor's performance as the central source of communication and information could be improved.

The finished project

The project in question finished on time and within the budget. The relationships between the client and the main contractor and between the main contractor and subcontractor still continue today, with all parties working together on other contracts. The lessons learnt on this project have been used on other projects and the main contractor is working on establishing strategic partnering relationships both up and down the supply chain.

Conclusions

This chapter has sought to identify what benefits partnering can give to its stakeholders and what the potential risks and problems of adopting partnering are.

The chapter began with an overview of the development of partnering and progressed to a brief discussion of the definitions given to partnering and how it is categorised. The chapter then showed that partnering had numerous benefits both tangible and intangible for all the partnering stakeholders. These benefits can be seen to fit under the following general headings:

- improved contractual situation;
- improved communication and information flow;
- increased understanding;
- improved efficiency of resources;
- improved financial position;
- improved quality.

The potential risks and problems of partnering were then identified. The most significant of these were: commitment; changing company culture; and use of partnering as a marketing ploy in order to gain contracts. Moreover, empirical data showed that there is a premium to pay when using partnering – there is a likelihood that tendering costs will increase.

Overall it can be seen that partnering can assist in the procurement of buildings; however, it must not be seen as panacea to all problems. Partnering requires for many a mindset change for it to be successfully utilised. The big test for all those who wish to use partnering, and who are involved in it, is the urge to revert back to the traditional approaches when old problems arise.

References

Abudayyeh, O. (1994) 'Partnering: a team building approach to quality construction management', *Journal of Management in Engineering* 10(6): 26–9.

AGC (1991) *Partnering: a concept for success*, Washington, DC: Associated General Contractors of America.

CIBSE (1995) *Partnering: altogether now*, London: Chartered Institute of Building Services Engineers, November, 31–4.

CIDA (1993) *Partnering: a strategy for excellence*, Construction Industry Development Agency, Australia.

Construction Industry Institute (CII) (1991), In Search of Partnering Excellence, Special Publication, 17–1.

Cook, E.L. and Hancher, D.E. (1990) 'Partnering: contracting for the future', *Journal of Management in Engineering* 6(4): 431–46.

Cowan, C. and Gray, C.E. (1992) 'Project partnering', *Project Management Journal* **XXII**(4): 5–11.

Cowan, C., Gray, C. and Larson, E. (1992) 'Project partnering', *Project Management Journal* **XXII**(4): 5–11.

Crowley, L.G. and Karim, M.A. (1995) 'Conceptual model of partnering', *Journal of Management in Engineering* **11**(5): 33–9.

Harbuck, H.F., Basham, D.L., and Buhts, R.E. (1994) 'Partnering paradigm', *Journal of Management in Engineering* **10**(1): 23–7.

Hellard, R.B. (1995) *Project partnering: principle and practice*, London: Thomas Telford.

Infante, J. (1995) 'The relative merits of term partnering and project specific partnering', *Construction Productivity Workshop Report 18: Project Specific Partnering, Can the Benefits be Realised*, European Construction Institute, Loughborough University, Loughborough, November, 4–6.

Kubal, M.T. (1994) *Engineering quality in construction: partnering and TQM*, New York: McGraw-Hill.

Loraine, R. (1994) 'Project specific partnering', *Engineering, Construction and Architectural Management* **1**(1): 5–16.

Maloney, W.F. (1997) 'Improvement example from the 1990s: Increased use of design–build as a project delivery system', *Transfer of construction management best practice between different cultures: CIB Proceedings, W65 Organisation and Management of Construction*, Publication 205, Montreal; London: E & FN Spon, 183–7.

Matthews, J., Tyler, A. and Thorpe, A. (1996) 'Pre construction project partnering: developing the process', *Engineering Construction and Architectural Management* **3**(1/2): 117–31.

Matthews, J. (1996) *A project partnering approach to the main contractor–sub contractor relationship*, unpublished PhD thesis, Loughborough University, Loughborough.

Moore, C., Mosley, D. and Slagle, M. (1992) 'Partnering: guidelines for win–win project management', *Project Management Journal* **XXIII**(1): 18–21.

Mosley, D., Moore, C., Slale, M. and Burns, D. (1991) 'Partnering in the construction industry: win–win strategic management in action', *National Productivity Review*, (Summer): 319–25.

NEDO (1991) *Partnering: contracting without conflict*, London: National Economic Development Office.

Pakora, J. and Hastings, C. (1995) 'Building partnerships: teamworking and alliances in the construction industry', Construction Paper, 54, Chartered Institute of Building, Englemere, Ascot.

Plavsic, A. (1994) Preventing fraud while partnering, *Buildup* (December): 6.

Porter, S. (1996) 'Partnering 2 – Some practical matters', *Construction Law Journal* (January): 178–80.

RCF (1995) 'Trusting the team: the best practice guide to partnering in construction', Center for Strategic Studies in Construction, Reading Construction Forum, Reading.

Sanders, S.R. and Moore, M.M. (1992) 'Perceptions of partnering in the public sector', *Project Management Journal* **XXII**(4): 13–19.

Schultzel, H.J. and Unruh, V.P. (1996) *Successful partnering*, New York: John Wiley.

Stevenson, R.J. (1996) *Project partnering for the design and construction industry*, Chicester, Sussex: Wiley Interscience.

Uher, T. (1994) *Partnering in construction*, School of Building, University of New South Wales, Kensington, NSW.

Walter, M. (1998) 'The essential accessory', *Construction Manager* **4**(1): 16–17.

Weston, D.C. and Gibson, G.E. (1993) 'Partnering – project performance in US Army Corps of Engineers', *Journal of Management in Engineering* **9**(4): 410–25.

12 Selection criteria

Steve Rowlinson

Introduction

This chapter deals with the issue of selection of procurement systems. The presumption is that choice of an appropriate procurement system will lead to a successful project outcome; this makes an implicit assumption that the objective of a procurement system is to provide a successful project, however success may be defined. The preceding chapters have dealt with many issues relating to the procurement system process and it is the intention in this chapter to draw together these ideas and concepts and present an outline methodology for the selection process. The philosophy adopted is that of a contingency approach: by this it is meant that there are a number of alternative routes available from the beginning of the process to the end which are able to satisfy the objectives of the user and that a range of variables exist which, depending on their value, have an effect on the system as a whole.

It is essential that during the selection process the objectives of the client be stated explicitly in order that appropriate success measures can be defined. These objectives must provide the starting point for the selection process and, in the context of this book, all of the issues discussed in the preceding chapters must be addressed if these objectives are to be clearly defined. It is accepted, however, that in certain circumstances it may be impossible at the outset clearly to define a number of key issues. In such a situation it must be borne in mind that the selection process can only be a satisficing process rather than providing a definitive answer to the procurement system question.

This chapter is organised as follows. The basis for choice is discussed, drawing on previous research into issues such as client criteria for project performance, the profiles of different clients and the nature of procurement systems. The impact of different project types and technologies will be addressed. Who chooses the procurement system is the second item on this agenda. Traditional views and the impact of information technology (IT) on this will be addressed. The third key element in this chapter is the context in which the procurement issue is addressed. The question of whether or not the choice of procurement system can be made in an objective manner is discussed. The impact of culture on wider-ranging issues such as developmental aims and emergent management and economic issues, amongst these partnering and strategic alliances, will be presented. The choice of criteria to use in the selection of procurement systems will be discussed with reference to past published work, and a new basis for selection, accounting for changes in the process, will be addressed. The chapter finishes with a discussion of a new paradigm based on a strategic view of the land-conversion process, which is put into a process-based context. Many organisational driving forces such as the impact of IT, politics and economics will be addressed in this discussion. The chapter will end with a checklist for procurement system selection and some afterthoughts in terms of the suitability of existing selection criteria in a global context.

The basis for choice of a procurement style

The conventional view of selection criteria is that these criteria should be based around the concepts of time, cost and quality. Such an approach can be traced back to the early reports in the UK on procurement system performance (such as Banwell 1964; Wood 1975). Research in the 1970s and 1980s focused on these criteria (for example, see Sidwell 1984). Such criteria were used in the National Economic Development Office (NEDO) reports, 'Faster building for industry' (1983) and 'Faster building for commerce' (1988). Many client guides have been produced to assist in procurement system selection. Of these guides, perhaps the most comprehensive was 'Thinking about building', produced by NEDO (1985). This guide produced a selection criteria chart (ibid.: 6, 7) which had a list of nine separate criteria by which the client was expected to set priorities for its construction project:

- Time: is early project completion required?
- Certainty of time: is project completion on time important?
- Certainty of cost: is a firm price needed before any commitment to construction is given?
- Price competition: is the selection of the construction team by price competition important?
- Flexibility: are variations necessary after work has begun on-site?
- Complexity: does the building need to be highly specialised, technologically advanced or highly serviced?
- Quality: is high quality of the project, in terms of material and workmanship and design concept, important?
- Responsibility:
  - is single-point responsibility to you, after the briefing stage, desired?
  - is direct professional responsibility to you from the designers and cost consultants desired?

- Risk: is the transfer of the risk of cost and time slippage from you important?

These criteria have formed the basis of much subsequent research into this problem. In addition to cost, time and quality other criteria have been included such as change orders and division of responsibility and risk. This approach in fact determines a contract strategy rather than a procurement system, as argued in Chapter 2, but this issue will be addressed as this chapter develops.

More complex models

The problem being addressed is actually more complex than the mere determination of selection criteria to be matched to the various contract strategies. Liu and Walker (1998) opine that the perception of the outcome of projects is a complex problem, which raises a whole range of issues. They state that the evaluation of project outcomes is a derivation of project goals, participants' behaviour and the performance of project organisations. They argue that each person's perception adds its own dimensions to the evaluation of the project outcome and they construct a model with two levels of outcome. The first-level outcome is project success but this is linked to the second-level outcome, which is participant satisfaction. (This concept was measured rather crudely by previous researchers such as Sidwell 1984; Rowlinson 1988). The valence of the first is dependent on the instrumentality of the second [Liu, (1995) uses Vroom's (1964) model as one of the starting points for her model]. They identify factors of influence as being project complexity, reward, self-efficacy, goals and environmental variables. This analysis is based upon the behaviour–performance–outcome (BPO) cycle derived from industrial and organisational psychology.

Client and industry profiles

A number of writers have focused on the characteristics of the client and the impact of these characteristics on the contract strategy or procurement system. Sidwell (1984) and I (Rowlinson 1988) have operationalised these characteristics, and Masterman (1992) used the following typology to characterise clients:

public	private
experienced	inexperienced
primary constructor	secondary constructor

The assumption made by many writers is that a contract strategy is suited to a particular client type and that certain clients are more capable of dealing with certain strategies than are others. As a consequence, it is assumed that experience and expertise are a moderating factor on the performance of the strategy chosen. This would fit in well with the views expressed by Liu and Walker (1998). However, the alternative view is often put that only certain types of clients are able to utilise a particular contract stategy. The criticism of this viewpoint is that it takes a narrow view of contract strategy and procurement systems and does not take into account the fact that procurement systems can be adaptive organisations: all organisations involved in the procurement process can be learning organisations (see Chapter 5 for a discussion of organisational learning).

Hence, one returns to the views expressed by Green and Lenard in Chapter 3 concerning the complexity of the client organisation, and the concept of project success can then be rationalised using the model of Liu and Walker (1998). Thus, it is an important precept in this book to view project outcomes from a multi-organisational persepctive. As a consequence it is important to incorporate this multidimensionality into any methodology which seeks to select a procurement system. A deterministic view might be directed towards resolving this plurality by means of quantitative analysis that could be seen to reduce the conflict of objectives within the organisation. However, this runs the risk of solving a different problem from the one posed by the construction project and its participants. In essence, this approach simplifies the problem to an unacceptably naive level, with the consequence that the solution will lead to unacceptable outcomes. This leads back to the point made in Chapter 2; the construction industry is impatient of the complexity of the client body with which it deals.

Contract strategies

Attempts have been made to produce sophisticated and deterministic systems for contract strategy selection. These systems have centred on using criteria derived from the operational research and the construction management fields of study and applying these to the contract strategy selection process by means of the use of a sample of problems. Hence, these systems have been based on a review of theoretical knowledge and practical experience. An example of one such system is that developed by Skitmore and Marsden (1988) in the UK and, subsequently, a similar system was produced by Love et al. (1998) in Australia. The basis for both systems was the 1985 NEDO document that provided the criteria for the selection process. However, the weightings of the selection criteria adopted were derived from a questionnaire survey in both instances and this survey dealt with responses from a range of respondents within the construction industry. Whilst having some usefulness as a first guess at which contract strategy could be adopted, the fact remains that these systems take into account only a limited set of criteria in developing a solution to any procurement problem. Such an approach has three problems:

- there are whole ranges of alternative criteria which are not addressed by such systems;
- the range of procurement system alternatives provided are rather naive and are certainly not an all-encompassing set of procurement options;
- such systems are very much country-dependent and are also dependent on the particular time at which they are developed – what is acceptable in terms of criteria and performance in Australia in 1997 may be totally irrelevant to South Africa in 1999.

Selection criteria

This chapter attempts to address these weaknesses and to identify a set of issues rather than criteria which need to be addressed in developing a selection methodology for procurement systems, a wider issue than contract strategy.

Projects

A number of writers have dealt with the nature of the construction project and have included this as an issue to be addressed in the selection of a procurement system. The rationale behind this approach is that different projects will have varying degrees of complexity associated with them and so the contract strategy and procurement system must be fitted to this complexity and so to the particular project. The often-cited example is the use of management systems for complex projects such as hospitals. Undoubtedly, there is some basis for such an approach but the consideration purely of project characteristics is not a sufficient basis for the choice of procurement system. The ability to use experienced staff to act as the client project manager can moderate the effect of project complexity. Furthermore, complexity can be viewed as an organisational trait or it can be seen as a technical issue.

Technology

The technology involved in a project may well require special skills and training for the employees on the project. In this regard the procurement system cannot make up for the lack of technical ability within the project team. There is no substitute for experience, and no matter how well constructed a procurement system might be it will not achieve its objectives if it does not have within it competent and experienced personnel. Thus, a major consideration in procurement system selection must be the availability of sufficiently experienced and skilled staff in technical as well as managerial areas. Procurement systems are essentially about the management of projects but the technical problems associated with many construction projects cannot be ignored in devising the procurement system to be adopted.

As pointed out in Chapter 10, the use of appropriate technologies may well determine the procurement system adopted.

The environment

The selection process is an open system which receives information from its environment, transforms this and returns it as output to the environment. It is very important to know what forces from the environment drive the system and how these forces might change during the duration of the project life cycle. Walker (1996) states the forces are political, economic, legal, technological and sociological, and it is essential that the forces which may affect the project are identified prior to identifying the selection criteria. A model of the selection process is shown in Figure 12.1.

In-house capability

A key question to be asked at project inception is 'Can the client do this itself?' If the answer is no then it is essential that advice is sought, as:

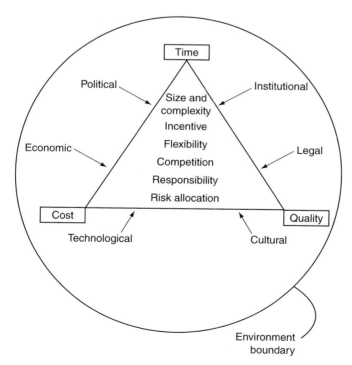

Figure 12.1 Forces acting on a project

In essence procurement advice can have an equally dramatic impact on a project's performance when measured in terms of time and cost as can design and engineering advice on the project performance when measured in terms of function and quality.

(Yates *et al.* 1991: 229)

So, where should this advice come from?

Choice of procurement system

Undoubtedly, the source of advice to the client organisation will have a major bearing on what procurement system is chosen. The construction industry has been widely criticised for poor performance in advising clients as to how to go about procuring construction projects. Wood (1975) stated that no other industry was so preoccupied with the relationships within the project team rather than with dealing with the client body. In some countries the profession of quantity surveying has taken upon itself the role of client adviser. One of the problems with this particular issue is that no one profession has within its charter the essential advisory role in this strategic decision process. In fact, there is a trend for external consultants, such as management consultancy firms and accountancy firms, to take on the role of procurement adviser. The construction industry thus risks losing to other industries a key role which it should be playing. It may well be that the construction industry as a whole has to become more generalist in its approach so as to be able to act as a client adviser.

A broker

Given that each profession in the construction industry has its own vested interests in advising clients it is perhaps necessary that some brokerage service is provided so that the client might perceive that it is receiving unbiased advice. Currently, such a broker does not exist. A number of attempts have been made to provide unbiased, deterministic advice to the client by means of electronic contract strategy selection systems. One such example was the ELSIE program (Turner 1990) developed at Salford University and used to advise clients on, among other things, which contract strategy to choose. The name ELSIE was derived from the abbreviation LC (lead consultant). The objective of the system was to give immediate advice to a prospective client on the design and cost planning of a building and, of interest here, to advise on the suitability of a limited number of procurement routes. The procurement route selected aimed to suit the client and its team. This system was an interesting attempt at automating the contract strategy selection process.

The previous mentioned systems devised by Skitmore and Marsden (1988; Love *et al.* 1998) also attempted to do this in a more transparent way.

One of the problems with computer-based systems is that they often appear to work as black boxes, and the logic behind the advice given is not readily apparent. However, the introduction of knowledge-based systems has allowed the incorporation of more explicit reasoning into some systems, and Alshawi's team at Salford University (Construct IT 1996) has produced the SPACE package through which clients can visualise the project in three dimensions and can check whether the design meets their requirements or budget; it also provides some advice on procurement route. SPACE therefore makes 'what if' scenarios quick and easy to carry out, and this is a facility which could be very useful for some clients in devising their contract strategy. To see the impact, in cost and time and quality, of changes of mind and how different contract strategies facilitate or hinder these changes is a very useful experience for clients. By seeing consequences of actions in 'what if' scenarios the black-box paradigm tends to disappear. When linked to visualisation technologies the process can almost mimic reality.

An adviser

These systems ostensibly fulfil the role of independent adviser to the client. However, it is obvious that such systems have a bias towards the opinions of the system developer. This is inevitable as there are no objective criteria for choosing contract strategies or procurement systems. The reason for this is that the choice is a strategic decision and can be made only by assessing and interpreting data, but these data are not all objective. Before they can be used in the decision-making process they must be assessed by the team of project participants and each will apply its own weighting to the data. Thus the methodology adopted by the adviser can only be a subjective approach which is informed by hard data on past performance of different strategies (see Chapter 10) and a range of semi-objective or subjective criteria derived from the project team.

Open approaches

Many of the systems which have been put forward for procurement selection are based on the black-box approach. Love *et al.* (1998) have attempted to make their selection system more transparent, but their system is firmly based on statistical analysis of past performance. A problem arises with this approach as, although the statistical analysis may give an indication of what is current practice, it does not necessarily indicate good practice. Given the fact that the T40 report (ABRGC 1994) in Australia and the Latham (1994) report in the UK have both indicated that a 30–40 per cent reduction in construction project time and similar reductions in costs are possible then it would seem that any analysis which has been based on past assessments of 'current practice' is flawed; past performance is recognised as being below par. Hence, if these systems set out to provide an optimal solution to the procurement problem then they will fail; due regard must be paid to bringing together the best aspects of practice with proven management theory.

Whether an optimal solution exists is a matter for debate; contingency theory would suggest that a number of solutions are available which will get the project from start to finish and still satisfy many of the participants' goals – satisfying all the goals in a complex organisation is not a likely outcome in most situations. In the real world one can argue that satisficing and suboptimal solutions are legitimate and sensible goals, particularly in the turbulent political, economic and social environments of the new millennium.

The context of procurement systems

In this section the context in which a procurement system is chosen and the context in which it operates are investigated.

Determinism

Is it possible to prescribe definitively the procurement system for any project? Is there a body of knowledge which enables us to say in any given situation that procurement system X is the best? The answer to these questions is currently no. Procurement systems operate in a complex project environment through a host of organisations with conflicting objectives. Hence, what is best for one organisation may not be the best for the others.

In the previous section the work of a number of writers was discussed in this context. Each of these had attempted to present a deterministic approach to the choice of contract strategy. Each of these had been successful to a greater or lesser extent. However, each accepted that the output from their system was in fact purely advisory and should be reviewed by the parties using the system.

Objectivity

One of the concepts on which the procurement system advisers have been based is that of objectivity. By this we mean that the advice given by the system can be seen to be unbiased and representative of the best knowledge available at the time. In an industry such as the construction industry where building types are constantly changing and the fortunes of contractors wax and wane any attempt to be totally objective in advice given is obviously a difficult task.

The problem being addressed is very complex and needs to have a detailed methodology provided for the client in search of an appropriate procurement system. It may well be that a contract strategy advisory system, such as ELSIE, SPACE or the Skitmore models, can be part of this procurement system methodology, but it will only be a part of the system and not the whole system. Additionally, it must be recognised that some of the decisions made will be based on subjective rather than on objective criteria. Although people plan to run businesses and organisations in a rational manner they also have to deal with other people in doing so. As a consequence it must be accepted that subjective judgments and non-objective criteria will be used in decision-making. This is, in fact, implicit in the work of Love *et al.* (1988) in which they indicate that the advice given by their system is only indicative and must be further explored. What is presented below is a discussion of the variables which need to be considered in the selection of a procurement system. Implicit within the discussion is a key principle that the selection and management of the procurement system is an ongoing process and does not finish once a set of organisations have been chosen and structured to facilitate the project. Although the key decisions made are strategic in nature and must be taken at the very outset of the project these decisions still have to be implemented, monitored and supplemented as the process continues. Hence, the procurement system has a life cycle which must be carefully monitored. It is thus necessary to consider what makes a procurement system successful and this issue is discussed with reference to research into critical success factors for projects.

A contingency approach

The approach adopted in dealing with these issues is a contingency approach. This approach deals with the reality of project management in that it accepts that there are many different routes leading to the same, or similar, outcomes and that standard approaches to projects are often not adopted fully owing to time, resource, political and other constraints. For example, the supposition that the brief provided by a client is a final and complete brief is often invalid. Hence, one must be able to adapt one's approach to suit the circumstances of the individual project, client organisation and project organisation. Many factors can be identified which have a bearing on the success of a procurement system and some of these factors, those considered most important, are dealt with in the sections below. Certain factors are environmental by nature whereas others are very much project-organisation-specific. One factor which has received considerable attention recently is the concept of culture.

Culture

The issue of culture in procurement systems has been dealt with in detail in Chapter 6 but the issue will be reviewed here from the perspective of the impact of culture on global construction organisations and, hence, its strategic impact on the development of procurement systems. Hall and

Jaggar (1997a) seek to illustrate some of the ways in which cultural differences can affect the procurement of international construction projects, examining the ways in which an organisation can respond to these cultural differences. They suggest that in light of recent international trading conditions, coupled with an ongoing trend towards globalisation, architectural, engineering and construction (AEC) enterprises wishing to compete successfully in the future will have to adopt a more strategic approach to their business and allow for a global dimension, stressing the importance of the cultural dimension to international business management. A primary cause of loss of international competitiveness is poor appreciation and understanding of cultural differences. They contend that recent literature relating to the impact of cultural diversity on the management of construction activity in the international environment is lacking in a number of key areas:

- it rarely attempts to put the impact of culture into a theoretical context;
- there is little empirical support for many of the observations made;
- strategic implications are only superficially explored and there is little attempt to fit those strategic implications into a company's overall strategic framework.

Hall and Jaggar state that cultural differences affect every stage of construction; for example, when a designer from one society or culture is designing a building for a different society, a special understanding of the end users and implications of their culture is required. As far as skills required for carrying out work on a building for a different society are concerned, they highlight from their research that company executives rank communication skills as most important for managers in a position of international responsibility, followed by leadership, interpersonal skills and flexibility. Functional and technical skills received a lower rating.

Hall and Jagger state that the synergistic approach is a rare response to a culturally diverse environment and occurs only when organisation members recognise the concept of culture as leading both to advantages and disadvantages. Members of a synergistic organisation believe that 'Our way and their way differ, but neither is inherently superior to the other'. If members of the organisation see the impacts of cultural diversity as having positive and negative effects within and upon the synergistic organisation then their strategy is to manage the impacts of cultural diversity rather than to manage the diversity, thus minimising problems and maximising potential advantages. Synergistic organisations train staff to recognise cultural differences and to use those differences to create competitive advantages for their organisation.

To allow for cultural differences it is suggested that a more holistic approach to training and education is required, encompassing the cultural dimension. This would provide managers at all levels with an awareness of the need to develop a collection of cross-cultural management skills. In conclusion, Hall and Jaggar insist that the global construction industry needs increasingly to focus upon other aspects of business strategy, such as on a greater understanding and appreciation of different cultures, if it is to survive as a world force into the next millennium. If this is integrated into procurement arrangements then benefits would be experienced throughout the organisation. If this issue is not addressed then companies could become increasingly marginalised and their markets could become increasingly parochial. Attention to cultural differences should be an essential element of the strategic procurement system at an international level. By embracing cultural management techniques outside of their domestic environments AEC enterprises could enjoy benefits such as partnering and improved collaborative arrangements. This is important in an international context as projects tend to be beyond the means of individual construction companies because those projects are of greater size and increased complexity. Build–operate–transfer (BOT) projects are increasingly popular in newly industrialising countries (NICs), where the developer has regular contact with the government and where an appreciation and management of cultural differences is vital to project success. The goal of providing transfer of management and technical expertise is facilitated by a successful cultural management programme.

In a second paper, Hall and Jaggar (1997b) extend their research into competition and cultural diversity in the international construction industry. They recognise that cultural diversity has long been recognised as a major determinant of an enterprise's success when operating in the international market. This is reflected in the extensive training of personnel in cross-cultural issues and in company policies when working overseas. They explore the issues involved in the globalisation of the AEC industry with an emphasis on the understanding and appreciation of different cultures as a key feature of the competitive strategy of such an enterprise.

Hall and Jaggar draw upon the fact that within the World Trade Organisation (WTO) there is a definite move towards globalisation of world trade, and internationalisation of industry in general can be said to be part of an ongoing tendency towards globalisation. Just as various manufacturing, media and service industries have become dominated by a small number of globally orientated organisations operating in all parts of the world, the same is occurring in the construction industry. As countries develop, the level of construction, within that economy, diminishes. The market for construction services is far from restricted to countries from the industrialised world as industrialising countries already play an important part in the global market and it is important to think of these countries as part of the international construction system.

Hall and Jaggar suggest that global AEC sector enterprises can no longer rely on traditional competitive advantages such as superior technical expertise and historical market conditions; rather, they will have to adopt a more strategic approach, focusing on aspects of management which have previously received little attention. A key aspect of globalisation of construction activity is the increasing tendency for construction enterprises to operate strategically across national borders. The creation of multinational consortia and joint ventures enables the pooling of technical expertise, reduces risk exposure and circumvents trading barriers. They conclude that industry was dominated by companies from the industrialised world that sought to work in the industrialising world. Using their position of dominance, such organisations could impose design and construction management techniques upon the host countries, with little regard for the cultural requirements and characteristics of those countries. Owing to the nature of the changes in global competition within the construction industry, global AEC enterprises are now encountering competition from industrialising regions and they need to adopt a more strategic approach to the conduct of their businesses if they wish to thrive in this dynamic global environment. The cultural dimension will be a key competitive feature of such a strategic approach and will have implications for all aspects of international construction procurement arrangements in the future.

Gilham and Cooper add to this opinion:

> In order to increase the chances of achieving sustainable construction, we need to recognise that the role of individual, group and national cultural characteristics deserves much more time, effort and understanding than we currently afford them. For successful developments, especially in the context of combating unsustainability, will need to consider the values, aspirations and motives of a broader range of players than traditionally has been necessary in construction procurement. These players will have to stretch beyond the individual project/client-base, as well as beyond the supply side of the individual design/construction team. Eventually, to match the principles underlying sustainable development, our construction procurement protocols will have to be capable of accommodating other (currently unrepresented) stakeholders, especially those involved in processes for regulating the production, maintenance and replacement of the built environment.
>
> (*idem* 1998: 256)

Critical success factors

The search for critical success factors (CSFs) has been far-reaching in the construction industry. Many have proposed a whole range of factors which they deem to be critically important to the success of construction projects. Mohsini and Davidson (1986) investigated the effect of organisation structure and conflict on project performance. Both Ireland (1983) and Walker D. (1994) investigated the determinants of construction project time performance and a very comprehensive attempt at this in a non-construction environment was undertaken by Pinto and Kharbanda and is reported in *Successful project managers: leading your team to success* (1995: 74). They reported 10 key factors:

- communication;
- project mission;
- top management support;
- project schedule or plan;
- client consultation;
- personnel recruitment, selection and training;
- technical tasks;
- client acceptance;
- monitoring and feedback;
- troubleshooting.

   In addition to these factors might also be added the following, which are based on the analysis by Walker and Kalinowski (1994):

- an understanding of team interdependence;
- cohesion;
- trust;
- enthusiasm.

In addition, the following factors can be derived from Mohsini and Davidson (1986) and from Walker (1994):

- conflict and its resolution;
- management of temporary multi-organisations;
- construction team flexibility;
- construction team control and administration skills.

Communication

Communication channels should be designed as part of the organisational structure of the project team and it should be the responsibility of all to ensure that communication channels are kept open and that high-quality information is transmitted in a timely manner. The client, the project team and top management must all be involved in this process.

The project mission

It is important that a clear project mission be accepted and understood by all team members so that a sense of the overall project goals is apparent. For example, such an approach involved the vertical fast-tracking up and down the organisational structure in the Hong Kong Convention and

Exhibition Center project. In any project, the attitudes, political orientation and leadership styles of the key players both in the client and in the construction organisations play a significant role in project success.

Top management support

In the project environment top management support for authority, direction and goal setting is crucially important. If resistance at top management level, within any of the participating organisations, exists then project progression will be severely inhibited. The key role that top management plays is in the provision of resources.

Project schedule and plans

Planning at all stages of a project is an essential prerequisite for success. Walker (1994) identified the approach and attitude to planning as a key element in the successful reduction of construction project time. The use of a measurement system when considering planning which allows for the judging of performance is a key success factor.

Client consultation

Client consultation has been identified by Green and Lenard in Chapter 3 as a prerequisite for project success. The degree of client involvement in project realisation is an issue which is becoming increasingly important as clients and projects become more sophisticated. Getting close to the client and understanding the client business is an essential production as well marketing factor.

Personnel recruitment, selection and training

Recruitment, selection and training of project personnel is becoming evermore important for project success. The key issues are not only of appointing those with technical skills but also of appointing those who can enhance team performance and compatability. People are a situational variable which can strongly influence the performance of the project team, and a framework within which selection of team members with appropriate skills and temperament can be enlisted to the team is required.

Technical tasks

As well as employing appropriately skilled personnel it is important that adequate technology with appropriate technical back-up is provided. In a global project environment one cannot rely on a full range of technical skills being readily available, and decisions have to be made as to whether the technical specification should be changed or the necessary expertise brought in.

Client acceptance

At the final stage of the project the client must be in a position to take over and operate the constructed facility. The project team cannot just walk out of the door at this stage; it must be able to tutor the client through the operation phase and be prepared to conduct a thorough project review with the client in order to learn and document lessons learnt during the project process.

Monitoring and feedback

This is part of the project control process, and adequate resources should be allocated to this issue during all phases of the project. This is an essential element of the planning system and allows for the identification of potential problems, provides opportunity for their rectification and should be applied to the performance of the team and its members as well as to the technical aspects of the project.

Troubleshooting

Although it would be very good to have a monitoring system which kept the project free from surprises it is not often possible to be in such a happy position. Hence, it is essential that a mechanism is set up to deal with serious problems immediately should they arise. The reaction time must be short, and resources must be set aside for this task.

Understanding team interdependence

One can simply define interdependence as the degree of joint activity amongst team members. What is needed in these circumstances is the engendering of a philosophy of cooperation between groups and a high degree of coordination. Good communication channels are essential for this to happen and the differentiation can be put into a positive setting and enhanced by producing nightmare scenarios which can subsequently be used as the basis for contingency plans.

Cohesion and commitment

Cohesion is the strength of desire of all members of a team to remain a team. Commitment is the desire of an individual to remain within an organisation and to accept that organisation's goals. Cohesion can bring about success by helping the team to concentrate on its technical and managerial tasks without having to deal with a lack of commitment from members. Project managers need to attempt to improve the cohesion of their teams, and techniques such as brainstorming and value engineering can assist in this as they allow forward thinking and prediction of problems so that these problems do not come as a nasty shock. Cohesion and commitment are also cultural issues which need to be carefully addressed in multicultural teams.

Trust

Trust is a difficult concept to describe but has been put forward as 'the team's comfort level with each individual member'. Trust is manifested in the ability and willingness to address differences of opinion, values and attitudes and to deal with them accordingly. Cohesion and commitment are difficult to maintain if trust is lost. In Chinese cultures the concept *Guang Xi* (reciprocation, relationship building and bonds) assists in the building of trust, and the technique of partnering can also assist in bringing this about. However, ultimately the issue of trust is an individual-to-individual matter and so again must be carefully monitored by the project manager.

Enthusiasm

If one wishes to implement a project then energy and drive are key components in the project realisation, and enthusiasm helps to create these. This enthusiasm is a reflection of motivation, and motivation theorists indicate that tasks act as motivators as long as those performing the tasks feel they are able to achieve the goals set for the tasks. Hence, enthusiasm and motivation can be driven from the top of the organisation.

Conflict

Conflict is inevitable in a project environment where many specialists work together in attempting to devise a best solution. Conflict can be very productive if it leads to better solutions, but it must be managed carefully, with skill and tact. Conflict must not be allowed to spill over into personal issues, which can be highly destructive.

Managing temporary multi-organisations

The process of managing a temporary multi-organisation (TMO) is a complex one as the basic premise is that one is managing a coalition rather than a team. Hence, the overriding concern is to be able to deal with the conflicting objectives of the coalition of organisations whilst still managing the TMO in such a way that the project objectives are achieved. This requires a great deal of tact, diplomacy and skill in order to weld together the coalition for the benefit of the project.

Team flexibility

In construction projects, the ability to change and adapt to the changing project environment is an essential skill. An unbending approach to suggestions and new ideas stifles innovation, and the value management approach, where alternatives are brainstormed and assessed and implemented, is important in the project environment. This is an attitude of mind which must be engendered in the project participants.

Team control and administration skills

Projects are always developing and changing and in order to manage such a fluid process it is essential that the whole can be controlled and actions be taken at appropriate points in time. Thus a well-structured administrative system is essential – that is, one which can respond to change and pressure without becoming overloaded or reacting negatively whilst still ensuring that goals are being met.

Figure 12.2 indicates how these factors are ranked in a hierarchy of factors affecting procurement system success. It can be seen that the traditional view, which encompasses 'contract strategy' criteria, consists of a number of the subsystems shown in this project performance model.

Procurement systems are a combination of all of the subsystems of multi-organisations and as such are complex managerial problems. The model below (Figure 12.3) attempts to draw together some of the key management system variables in a way which illustrates their strong interdependence. The core of the model is leadership, motivation and commitment. However, all of these variables are affected by the organisation's structure, the organisational culture, the prevailing national cultures and the environment of the project. Hence, what is being presented here is a complex conceptual model of the way that a project management system works. This model is a first-level conceptual model and is capable of being expanded at a number of levels for any of the variables shown in it. The essence of the model is that project management systems are a complex organisational issue and so past attempts to model procurement systems by simple contract strategy variables have been too impatient of the system complexity. There is no simple solution to the design of procurement systems but there are a whole series of key variables which need to be addressed, and these are listed later in the next section, on selection criteria.

Thus any critical success factors which have been identified need also to be reviewed in the light of environmental circumstances, including demand and supply factors. The framework within which this review should take place is the classical systems theory framework of:

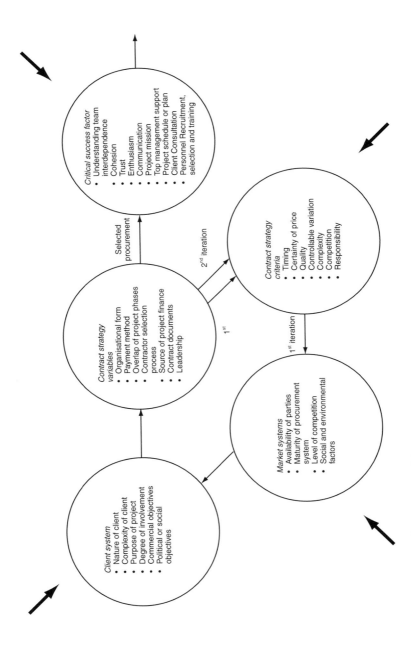

Figure 12.2 A project performance model

Critical success factor
- Understanding team interdependence
- Cohesion
- Trust
- Enthusiasm
- Communication
- Project mission
- Top management support
- Project schedule or plan
- Client Consultation
- Personnel Recruitment, selection and training

Selected procurement

2nd iteration

Contract strategy criteria
- Timing
- Certainty of price
- Quality
- Controllable variation
- Complexity
- Competition
- Responsibility

Contract strategy variables
- Organisational form
- Payment method
- Overlap of project phases
- Contractor selection process
- Source of project finance
- Contract documents
- Leadership

1st

1st iteration

Market systems
- Availability of parties
- Maturity of procurement system
- Level of competition
- Social and environmental factors

Client system
- Nature of client
- Complexity of client
- Purpose of project
- Degree of involvement
- Commercial objectives
- Political or social objectives

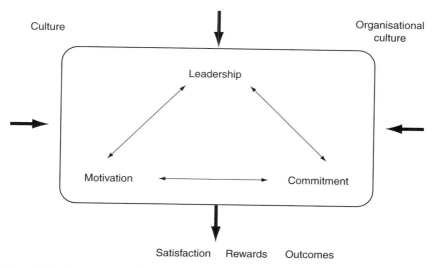

Culture   Organisational
                                                      culture

Figure 12.3 Organisational effectiveness

- political environment;
- economic environment;
- environmental issues;
- socio-technical system (including culture);
- organisational system;
- structural system.

Implicit within these factors will be a consideration of issues such as sustainability, environmental and developmental objectives, safety goals and corporate strategies such as partnering, strategic alliances and image. The point of this discussion of critical success factors is that there is a need to look beyond the simple selection of a contract strategy in developing an appropriate procurement system. This procurement system must be able to deal with the environmental forces acting on the project and its participants and be able to deal specifically with many of the technical and psychosocial aspects of project management. It is this aspect which is a weakness of the contract strategy advisory systems currently in use. Hence there is no easy solution to the selection of an appropriate procurement system. What is presented below is a checklist of important factors which reside within the procurement systems models and which need to be addressed and which, if dealt with in detail, will lead to the adoption of an appropriate procurement system for the project in prospect. These factors need to be monitored throughout the lifecycle of the project.

Selection criteria

The factors listed below are broken down into four broad categories which represent the strategic issues raised in the discussion above and are based on research and developments which have taken place over the past 20 years. Procurement system choice is a strategic decision and the factors listed below are currently relevant but will need to be periodically reviewed; not all factors will be applicable to each individual project and some may have more relevance (weighting) for some countries than others. With rapid advances in IT, areas such as data sharing, object-orientated programming, collaborative design and visualisation new paradigms for the design of the construction procurement process will emerge very soon. The factors listed below will still be relevant to these new paradigms.

Project performance factors

Project performance factors are those higher-lever prerequisites which any organisation must pay attention to getting right in order that a project continues to be managed to a succesful conclusion:

- culture;
- leadership;
- motivation;
- commitment.

Critical success factors

The critical success factors are those fundamental issues inherent in the project which must be maintained in order for teamworking to take place in an efficient and effective manner. They require day-to-day attention and operate throughout the life of a project:

- understanding of team interdependence;
- cohesion;
- trust;
- enthusiasm;
- communication;
- project mission;
- top management support;
- project schedule and plans;
- client consultation;
- personnel recruitment, selection and training;
- technical tasks;
- client acceptance;
- monitoring and feedback;

- troubleshooting;
- conflict resolution;
- team management.

Contract strategy variables

These are the key determinants, chosen at the outset of a project (by positive choice or default) which determine the structure within which the project will take place and within which the critical success factors operate:

- organisational form;
- payment method;
- overlap of project phases;
- contractor selection process;
- source of project finance;
- contract documents;
- leadership;
- authority and responsibility.

The environment

All of the factors listed above operate within the project environment, which is affected both by macro and by micro factors, such as:

- political environment;
- economic environment;
- environmental issues;
- socio-technical system;
- organisational system;
- structural system;
- goals and values system.

Structured methodology

Hence, a structured methodology for the development of the procurement system starts to become apparent. The initial consideration of contract strategy is appropriate only when considered in conjunction with those environmental forces which should determine the weighting applied to the various factors under the critical success factors; only then can we consider contract strategy. These environmental forces will impact directly on how the critical success factors are used and evaluated. Thus, at the earliest stage of project development the environmental forces will moderate the critical success factors. These factors will be the key determinants of the contract strategy. Once the contract strategy is chosen the project performance factors will come into play and the critical success factors will determine how these performance factors are managed. This process will be continuous; once the contract strategy has been adopted attention must be paid constantly to the critical success factors and the project performance factors. The initiator of the project must be willing and able to adapt these factors throughout the course of the project to ensure success.

References

ABRGC (1994) 'T40 construction research', Austrialian Building Research Grants Committee, Department of Industry, Technology and Regional Development, Canberra.

Banwell, G.H. (1964) *The placing and management of contracts for building and civil engineering work*, London: The Stationery Office.

Construct IT (1996) http: //www.construct-it.salford.ac.uk/pages/conferences/memb0598/docs/report

Gilham, A. and Cooper, I. (1998) 'Exploring the cultural dimensions of construction procurement – dealing with difference to achieve sustainable development', *Proceedings of CIB World Building Congress*, vol C, Gävle, Sweden, 246–58.

Hall, M.A. and Jaggar, D.M. (1997a) 'The international construction industry, competition and culturally diverse environments', *Procurement – a key to innovation, Proceedings of CIB W92 Symposium*, Montreal, Canada, 233–41.

Hall, M.A. and Jaggar, D.M. (1997b) 'Accommodating cultural differences in international construction procurement arrangements', *Procurement – a key to innovation, Proceedings of CIB W92 Symposium*, Montreal, Canada, 243–50.

Ireland, V. (1983) *The role of managerial actions in the cost time and quality performance of high-rise commercial building projects*, PhD thesis, University of Sydney, Sydney.

Latham, M. (1994) Constructing the team. *Joint review of procurement and contractual arrangements in the United Kingdom construction industry: final report*, London: The Stationery Office.

Liu, A.M.M. (1995) *Analysis of organizational structures in building projects*, PhD thesis, Department of Real Estate and Construction (previously Surveying), The University of Hong Kong, Hong Kong.

Liu, A.M.M. and Walker, A. (1998) 'Evaluation of project outcomes', *Construction Management and Economics*, 16: 209–19.

Love, P.E.D., Skitmore, M. and Earl, G. (1998) 'Selecting a suitable procurement method for a building project', *Construction Management and Economics* 16: 221–33.

Masterman, J.W.E. (1992) *An introduction to building procurement systems*, London: E and FN Spon.

Mohsini, R., and Davidson, C.H. (1986) 'Procurement, organisational design and building team performance: a study of inter-firm conflict', *Proceedings of the 10th Triennial Congress of the International Council for Building Research, Studies and Documentation*, vol. 8, Washington, DC, 3548–55.

NEDO (1983) *Faster building for industry*, National Economic Development Office, London: The Stationery Office.

NEDO (1985) 'Thinking about building: a successful business customer's guide to using the construction industry', National Economic Development Office, London.

NEDO (1988) 'Faster building for commerce', National Economic Development Office, Commercial Building Steering Group, London.

Pinto, J.K. and Kharbanda, O.P. (1995) *Successful project managers: leading your team to success*, New York: Van Nostrand Reinhold, 74.

Rowlinson, S. (1988) *An analysis of the factors affecting project performance in industrial building*, PhD thesis, Brunel University, Uxbridge, Middx.

Sidwell, A.C. (1984) 'The measurement of success of various organisational forms for construction projects', *Proceedings of CIB W95 4th International Symposium on Organisation and Management of Construction*, University of Waterloo, Waterloo, 283–90.

Skitmore, R.M. and Marsden, D.E. (1988) 'Which procurement system? Towards a universal procurement selection technique', *Construction Management and Economics* **6**: 71–89.

Turner, A.E. (1990) *Building procurement*, London: Macmillan Education.

Vroom, V.H. (1964) *Work and motivation*, New York: John Wiley.

Walker, A. (1996) *Project management in construction*, 3rd edn, Oxford: Basil Blackwell.

Walker, A. and Kalinowski, M. (1994) 'An anatomy of a Hong Kong project – organisation, environment and leadership', *Construction Management and Economics* **12**: 191–202.

Walker, D.H.T. (1994) *An investigation into factors that determine building construction time performance*, Ph.D thesis Department of Building and Construction Economics, Royal Melbourne Institute of Technology, Melbourne.

Wood, K.B. (1975) *The public client and the construction industries*, National Economic Development Office, London.

Yates, A., Dearle and Henderson (1991) 'Procurement and construction management', in P. Venmore-Rowland, P. Brandon and T. Mole (eds) *Investment, procurement and performance in construction*, London: E & FN Spon, 219–35.

Appendix 1
Procurement games – a learning process

The aim of providing these games or role plays is to allow both practitioners and teachers the opportunity to explore some of the key concepts in procurement systems by means of role play. By exploring issues in this 'safe' environment the users can be made aware of some of the issues in their own project to which they must pay serious attention. This will hopefully lead to better informed decisions.

A. The building game

The following is reproduced by kind permission of Professor David Langford of Strathclyde University, UK.

Description

This game aims to indicate to the players the difficulties involved in communicating construction details to the various members of the product team. Each player undertakes a distinct role in the process and communications with the architect are allowed only in written form. The game should help participants to understand how difficult it is to communicate design and construction needs by the conventional written forms. By making everyone aware of these difficulties it is possible that the participants will be more careful and thorough when communicating during the construction procurement process.

1. Objectives

The objective of this game is for the building team to replicate the model for the least cost in the shortest possible time.

2. Teams

In a team of 5 you will require persons to fill the roles of:

(a) architect

(b) quantity surveyor
(c) buyer
(d) builder
(e) contracts manager

3. Roles

(a) The architect:

The architect looks at the model and records on a sketch pad how he or she perceives its composition in terms of shape, colours and components. The sketch pad is the only allowable means of communication with the team. The architect is not allowed to touch the components or speak.

(b) The quantity surveyor:

From the architect's drawings, the quantity surveyor extracts the numbers of the various components required. He or she communicates this information to the buyer. The quantity surveyor is also required to keep up-to-date cost records. He or she will be responsible for preparing the final account.

(c) The buyer:

The quantity surveyor will report to the buyer that the building has x number of red components, y number of blue components etc. The buyer then fills in the proforma provided and delivers it to the builders merchant. The buyer must fill in the order *and* delivery times. The following delivery times are applicable.

Red	–	10 minutes
Black	–	6 minutes
Blue	–	Immediate
White	–	2 minutes
Yellow	–	15 minutes

(d) The builder:

The builder is responsible for the assembly of the components; only he or she may assemble the components.

(e) The project manager:

The project manager must manage the situation in such a way that the group achieve the objective of the game.

4. Constraints

(i) Only the architect may visit the model. The first visit to the model is free; for subsequent visits the team pay a fee of $10,000 per visit.

(ii) The architect may only communicate via the sketch pad.

(iii) Only the buyer may buy components.

(iv) Only the builder may assemble the components.

(v) No role-swapping is allowed. The site will immediately be brought to a halt by a demarcation dispute if any team member is observed fulfilling another member's role.

(vi) The contract should be completed in 40 minutes. Completion is considered from the time that the client accepts the building as complete.

(vii) Time penalty: each minute over the contract time of 40 minutes will add $5000 to project cost.

(viii) Costs: Each Component will cost the following:

Red	–	$70,000 + $2,000 delivery per load
Black	–	$100,000 + $2,000 delivery per load
Blue	–	$5,000 + $500 delivery per load
White	–	$20,000 + $1,000 delivery per load
Yellow	–	$120,000 + $3,000 delivery per load

5. Contract budgets

The job has been tendered for and obtained at $.

6. Nuisance factors

Various nuisance factors may be introduced during the game by the monitor. These may involve safety problems, industrial relations problems, materials shortages etc. The contracts manager is the only person allowed to negotiate with the monitor over these issues.

7. Completion of model

On completion of your model the monitor will compare the structure you have built with the original model and will have the discretion of advising teams on the accuracy of their structure. Incorrect assembly incurs the following penalties.

Each missing component	–	2 × component cost
Each component in wrong place or wrong colour	–	1 × component cost

8. *Team performance*

Each team will be given a statement of penalties from which penalty costs can be calculated. Team performance will be compared on profit/loss calculated on final account sheets.

9. *Attached documents*

Order forms
Final account sheet
Sketch pads

Forms to be handed to each team

Final Account

COMPANY NAME	
Total cost of materials purchased	$
Total delivery costs of materials	$
Extra costs incurred in site visits at $10,000 per visit	$
Liquidated & Ascertained damages due to late completion @ $500/minute	$
Penalty for incorrect assembly:	
... No. components missing @ × 2 cost	$
... No. components wrongly placed or in wrong colour @ cost	$
Actual Final Cost	$
Estimated Cost	$
PROFIT/LOSS	$

Materials Order Form

COMPANY NAME: . . . .

Brick Colour	No. reqd	Time ordered	Delivery time

Figure 1 MSc students at HKU

Figure 2 A typical project

B. Construction Procurement Simulation

The following is reproduced by kind permission of Robert Newcombe, The University of Reading, UK.

Description

This simulation aims to indicate to the participants the difficulties involved in determining the criteria for selection of a procurement form and putting together a comprehensive and consistent brief. It also shows how requirements can easily be misinterpreted. The simulation should help participants to understand how difficult it is to communicate procurement system priorities and criteria and also make it clear that choice of procurement system is an inexact science. As with the previous game, it should make everyone aware of the difficulties in communicating effectively so that participants will be more careful and thorough when communicating during their construction procurement process needs. As an aid to this simulation participants may wish to refer to 'Thinking About Building' as a starting point for this process – *Thinking About Building. A successful business customer's guide to using the construction industry*, 1985, Great Britain, National Economic Development Office.

Stage 1 (20 minutes)

Client brief

You are to act as one of the following types of client:

> (a) A **HIGH-TECH** company which needs a new laboratory to produce computer hardware.
> (b) A **LOCAL AUTHORITY** which intends to build a new Town Hall and Headquarters.
> (c) A national **FOOD RETAILER** which wants to build a new out-of-town supermarket but has professional services in-house.
> (d) An international **MANAGEMENT CONSULTANCY** firm which has bought a new speculatively-built office block in London which it intends to fit out as its corporate headquarters.

Develop a set of criteria as client for one of the above and establish priorities and a scenario.

Consultant's brief

Consultants should **separately** develop a set of general criteria that a client for a building project might have.

Assess the extent to which the various procurement paths will meet these criteria by developing a comparison matrix.

Stage 2 (4 × 10 minutes)

Consultants to interview each client in turn to determine their criteria and propose a procurement path most appropriate with reasons for choice.

Stage 3 (5 minutes each consultant)

Plenary review

Each consultant to present client criteria and chosen procurement path with **reasons**. Client groups to comment on the consultants' understanding of their criteria and the procurement paths selected.

Supplementary information for client groups

General

Each client group should develop a scenario for their project, e.g. location, value, duration, performance specification etc.

Client groups should only provide information asked for by the consultants; unsolicited information should not be given.

(A) HIGH-TECH:

The fast moving nature of the computer industry means that it is important to get into production as soon as possible so speed of construction and quality of finishes for a 'clean' environment are top priority. The cost of fitting out the laboratory is expected to exceed the building costs.

Assume any other information you need.

(B) LOCAL AUTHORITY:

To meet requirements for public accountability some form of competitive tendering is essential. The project budget must be strictly adhered to but an innovation design is seen as important to produce a prestigious local amenity.

Assume any other information you need.

(C) FOOD RETAILER:

Standard design requirements and specifications are used throughout the UK to ensure uniformity of appearance and quality between stores. A small department comprising architects, services engineers and quantity surveyors is based at Head Office. Certainty of completion time is critical for all projects undertaken.

Assume any other information you need.

(D) MANAGEMENT CONSULTANCY:

The reinforced concrete multi-storey office shell is to be fitted out to a high standard and to a very tight schedule. The client expects full involvement in the management of the project as they want to offer project management services to the construction industry in the near future and see this project as a learning opportunity.

Assume any other information you need.

Procurement selection matrix

		D & B		Traditional		Management	
		Direct	Comp	Sequ.	Accel.	MC	CM
High Tech	C1						
	C2						
	C3						
	C4						
Local Authority	C1						
	C2						
	C3						
	C4						
Food Retailer	C1						
	C2						
	C3						
	C4						
Management Consultancy	C1						
	C2						
	C3						
	C4						

Appendix B – Procurement selection matrix – results

		D & B		Traditional		Management	
		Direct	Comp	Sequ.	Accel.	MC	CM
High Tech	C1					S	SN N
	C2	SN N				S	
	C3	N				S	SN
	C4			N		S	SN
Local Authority	C1				S SN	N	
	C2			N	S SN		N
	C3				S SN		
	C4				S SN		N
Food Retailer	C1	S	N SN				
	C2		S SN			N	
	C3		S SN				N
	C4		S N SN				
Management Consultancy	C1				N		S SN
	C2				N		S SN
	C3					S	SN N
	C4						S SN N

Key: S = SOPHISTICATED   SN = SEMI-NAIVE   N = NAIVE

Index